Animals in Their Worlds

ANIMALS IN THEIR WORLDS

Helga Menzel-Tettenborn and Günter Radtke

Foreword by Peter Brazaitis,
New York Zoological Society

with 500 illustrations, 293 in colour

T&H

Thames and Hudson · London

Translated from the French
Le Monde Merveilleux des Animaux
by Laura Guggenheim

Printed in West Germany

ISBN 0 500 01105 2

Contents

Foreword

"When the last individual of a race of living things breathes no more, another heaven and another earth must pass before such a one can be again."

This simple, obvious truth was offered years ago by a man who spent a lifetime as a scientist and naturalist probing the secrets of earth's wild kingdoms—Dr. William Beebe, first curator of birds and late director of the New York Zoological Society's tropical research facility.

Today more than ever we find his words and the impact of their meaning coming to pass time and again, as one species after another is forced out of existence for eternity. Many animals, man among them, inhabit this planet in an intricate balance dependent upon the presence of each to insure the survival of all. The passing of even the seemingly most unimportant species creates a disturbance in the balance of nature like the effect of a pebble thrown into a quiet pond: each ripple radiates and affects the composure of the entire surface.

Wildlife around the world is in serious danger, with many hundreds of species on the brink of extinction and a great many more perilously close behind. Our governments, scientists, and zoological institutions can do only so much to bring about a reversal to these rapidly declining populations, and to insure the protection of others. The major portion of the responsibility rests squarely with the masses of ordinary people. We all must become conscious of the presence and living value of wildlife, and be motivated to insist on its continued survival.

Animals in Their Worlds, by Helga Menzel-Tettenborn and Günter Radtke, is a welcome addition to this international effort. The inspired collection of color and black and white photographs depicts each species discussed in a true moment of real life, with a care for detail that captures the mood of the scene. These are the worlds we must not lose.

In these pages we can join a pride of lions at home on the African veldt. A lioness uses the sparse brush to conceal her movements; a regal—and sleepy-eyed—male is caught at rest in the hot African sun. We see a lion family at play, and spot the lord of the plains in a tree, his potbelly hanging and a ridiculous look on his face. He may have just eaten; hyenas and vultures will feast on the leftovers from his meal. These are the all-important scavengers who keep the plains free of decomposing carrion. There are other plains dwellers: fleet-footed gazelles, antelope, and zebra. They live in the open, depending on their speed and alertness rather than shelter to keep them safe.

We are transported over the globe by way of exploration of eleven biomes or basic habitat situations. A shark prowls the blue-water reefs of a warm sea, while a bull gorilla stands guard over his family at home in the tropical forest treetops. The photographs speak for themselves: the accompanying text is rich in fact, yet presented in a style that will appeal to a broad spectrum of young and adult readers. The combined work reflects our entrance into this area of concern.

We have always looked upon wildlife and our natural resources as inexhaustible commodities to be used without reservation or foresight. Unfortunately, wild animals must not only withstand the pressures of direct hunting but must also bear the consequences of a lost habitat.

We wrap the hunt in the blanket of "sport," "control," or "commercial need." Because we have already eliminated the wild predators, in many instances these rationales are legitimate. The sport-hunting of deer in the United States, for example, not only helps to control their numbers and provide meat for the hunter, but also, through licensing, produces revenues that help finance conservation programs. Game management practices developed in these programs insure the continued presence of deer. But this chain of control is often necessary only because man has been too quick to "control" predators, without considering the total value of their presence. Then he finds himself artificially controlling the very species previously (and better) held in balance by their natural predators—and at lower cost.

But how can we justify the continued killing of animals such as whales for fertilizer, dog food, or oil no longer required in this day of modern technology? Most whales are no longer found in sufficient numbers to insure their survival. The whaling industry obviously requires whales to stay in operation; but the operation of the industry, together with other ecological disruptions, will mean the extinction of most whales and thus of whaling itself. The situation is not unique; our use of wildlife has frequently been to exploit until the animal is no longer available—and then go on to the use of another species. But today there are fewer other species to go on to.

Animals cannot adjust their manner of reproducing, feeding, or habitation with the ease that man can adjust his methods of predation. A polar bear still must live and hunt on the pack ice of the arctic; the female still gives birth to only one or two cubs at a time. Eskimos hunted the bear for fur and food, making the kill after a long and dangerous journey in small boats, sleds, or on foot, frequently with primitive short-range weapons. Today, hunters in light planes can cover hundreds of square miles in a single day and shoot more bears in a shorter period of time with longer range rifles. All too often the hunter is motivated by his desire for a trophy rug on the living room floor rather than need for the meat as food or the fur for warmth.

The cheetah or leopard coat, although no warmer, more durable, or more attractive than its less expensive man-made counterpart, may be the last word in fashion. It may also represent the entire cheetah or leopard population in an area of many square miles. A few fur coats on the streets of London or New York could mean a significant alteration in the ecology of a vast area. How? In simplest terms, the loss of the most effective predators could cause a population explosion the grazing animals of that particular region. The expansion of farm lands on to what might once have also been usable habitat could then compound the situation by forcing more animals to live in a smaller amount of space. This would mean an increase in demand on the environment that the taxed vegetation could not support. Once the vegetation is stripped away, the habitat is destroyed. Not only do the grazing animals suffer the end of a food supply, but the secondary citizens of the ecological community are deprived of nesting materials, associated food supplies, and cover. And what of the local peoples and the impact on them of a land devoid of grass and game?

It is difficult to believe that, at the present rate of decline, many species familiar to us now may become extinct creatures within a generation, largely because of the pressures of excessive commercial hunting and the loss of suitable habitat. Polar bears, leopards, cheetahs, tigers, most species of crocodilians, most whales, many birds of prey, gorillas and other great apes, and others too numerous to mention, are thus in danger. For some individual species it may already be too late: too few individuals may remain, and too little is known of their natural history to initiate meaningful breeding programs. Others, even if propagated under captive conditions, no longer have a native habitat to be returned to. Zoological institutions are doing their utmost to breed the most threatened species, those with little or no hope for survival in the wild. But it is only in the natural state that true survival for the vast majority of earth's wildlife exists. And this survival can be insured only through the intelligent mangement and use of wildlife, and the habitat in which it lives. It is up to you, the reader, informed and motivated by books such as *Animals in their Worlds,* to value and to care.

PETER BRAZAITIS
New York Zoological Society

1. Plains, Savannahs, Prairies

The Lion

He is strong—he has unyielding muscles, a backbone of steel, paws like hammers, and a compact muzzle with viselike jaws. He is cautious—his cold glance observes and measures. He is psychologically powerful—his roar freezes the heartbeat of the African night. He has the sun-colored mane of an animal-god. Because of these attributes he is Simba, King of the Animals and Lord of the Plains. African folk tales, Western fables, and popular imagery all make him the symbol of great courage and infallible wisdom.

Underneath this lion of legend, there is the real lion. He is as strong as reputed; not as brave as he seems, but far more unusual. Of all the large cats, he is the only one that roars, as though he needed an outlet for his excessive vitality; that thundering sound can be a rallying cry, a signal for attack, or a loud sigh of contentment. He is the only feline that lives communally: he plays, sleeps, hunts, and eats in groups. In the daytime, the pride dozes in the shade of reeds, in the hollow of a thicket in its territory. At dusk, it silently glides toward watering spots and, if hungry, hunts. The hunting technique is usually the same. Some of the lionesses, the males, and cubs hide, under cover, on the lookout. Other females leave them, and chase a herd of antelope or zebra (their favorite prey), or warthogs, or even buffalo. They drive some of the herd towards the hidden members of the pride, who then join in the chase. One of the animals finally pounces on the back of a single victim and kills it with one bite or a blow to the head with a paw, then drags it toward the rest of the group, which rushes for the spoils. But the pride never neglects its special rituals: the first to eat is the leader of the pride, who rules over the lionesses and who never hunts. He has the right to the choicest piece and majestically begins the feast. If his prerogatives are ever questioned, he defends them by hitting the upstart. Once the leader's hunger is satisfied, hierarchy and manners are ignored. There is a great deal of heavy breathing, roaring, and scuffling, and the weakest may go hungry.

In the Ice Age, lions were found throughout the continents of Europe and Asia. They hunted wild horses and bison. But the European lion, the "cave lion," which was even more impressive than his cousin of today, disappeared five thou-

▲ "To roar is to rule" seems to be the motto of this extravagantly maned male (*Panthera leo*). But the most powerful lion races have been exterminated, and specimens as handsome as this one are hard to find at large today.

◀ Life means hunting: this slender light-coated lioness and her cubs waited for the lion to drag the prey, a zebra, to them.

On the move: hunted and tracked down from era to era, from country to country, by climatic conditions and by men, lions today are confined to the African plains south of the Sahara and to the preserves of South Africa and India. They never live in deserts without water holes. ▶

The king dreams. When a lion is full, he is peaceful. Many lions climb trees: the fork of a branch seems an ideal spot for rest or meditation. ▶

▲ The cubs are playing and their mother is playing with them, rolling them in the tall grass, distributing slaps as well as caressing them, and if need be biting them to curb their excessive exuberance.

sand years ago. Then, as civilization arrived and the lions themselves became prey, the Greek and Near Eastern lions disappeared. The two most powerful races remaining became extinct with the killing of the last Cape lions and then, in 1892, with the killing of the last Barbary lion in Tunisia. The survivors, smaller and with less spectacular manes than their relations, are now limited to the open plains and bush of Africa and western India, where they are protected from extermination by law.

The lion can live to the age of forty, but in general rarely survives beyond thirteen. He reaches sexual maturity at the age of three, but three more years pass before he is considered an adult. The females, who have no mane and are more supple and agile than the males, give birth to litters of two or three cubs each after a gestation period of a hundred days. The expectant mother retires to a thicket, where the birth takes place. The cubs remain there until they gain some semblance of independence. When they are old enough, they climb into trees, where they can sleep in complete safety, curled up against their mother. The lion actively participates in the upbringing of the young. Together with the lioness, he watches over the cubs, feeds them, and joins in their games. He is the only feline that manifests the paternal instinct.

It is difficult to reconcile the ferocious bloodthirsty carnivore depicted in the old tales with the accounts of tourists in automobiles who see a seemingly harmless, lazy, and debonair lion begging for meat along the roads in the African national parks. It all depends on the circumstances. Normally, the lion kills only in order to feed himself or to avoid human beings. A small animal like the zorilla, a perfectly harmless type of African polecat, is capable of terrifying the lion. A park keeper once watched five adult lions wait for

▲Group hunting: lions are the only felines to hunt collectively. They like to hunt at dusk, most often attacking large mammals. Here they are chasing a herd of gnu, brave and irascible African antelope.

hours until a zorilla had eaten all it wanted of a zebra the lions had killed. The lions did not dare to approach their spoils until the zorilla had left. Occasionally there are waves of murders among young lions, just as there are waves of juvenile delinquency among human young. But these instances are quite rare. On the other hand, a hungry lion is a serious matter. A lion that is threatened, at bay, under fire, or wounded, or a lioness defending her cubs, is dangerous. And a lone lion—pushed aside by his pride because he has become old and sluggish—who can no longer hunt anything but small rodents, may turn into a dangerous man-killer.

In ancient Egypt and Assyria, lions—even adults—were trained for hunting. In modern times, lion cubs raised in captivity may enjoy excellent relations with their keepers. *The Lion*, a love story of a lion and a little girl, which earned Joseph Kessel the Goncourt Prize, could well have happened in real life. People familiar with Africa have similar stories to tell.

The fate of lions in zoos has greatly improved as the depressing cement cages are gradually being replaced by open, moated areas where the animals live in semi-freedom under conditions approaching those in their native savannahs and which, partially at least, preserve their natural life style. Lions are capable of reproducing normally in such environments.

Tête-à-tête: the king of the beasts is often very tender toward his own. It is easy to forget that he is not always this affectionate.
▼

▲ Rest: the sun is at its height, and the lions seek shade in which to nap. They stay on the plains where their prey are found, always avoiding the thick forest.

Antelopes and Gazelles

The lovely word "antelope" is originally Arabic, meaning "flower eyes." No name could better suit those timid fleeing creatures of the savannah, rapid as the wind, fugitive as a dream, and forever threatened. As they stand on their tall, slender legs, they are always prepared to flee: their enemy is ever present, and their life without rest.

Coming originally from Asia, they now

◀ Antelope are sociable animals and get along well with the zebra with whom they cross the vast plains of the dark continent.

Grant's gazelles (*Gazella granti*) are the most impressive of the antelopes and the fastest gazelle of all.
▼

inhabit all of Africa: savannahs, plains, damp forests, mountain and river valleys. No other mammal, with the possible exception of the monkey, has as many species. Hardly larger than a roe deer, the gazelle is the prototype of the antelope. It has a fine head standing on a long neck, horns attractively bent backward, and it is supple and graceful from the tips of its hooves to the points of its horns. It likes

The delicate, friendly giant eland often duel outside the mating season. They fight in a kneeling position, and one of the combatants often loses one of his peculiar double-curved horns. ▶

The Cape eland (*Taurotragus oryx*) carries impressive screw-shaped war gear—sometimes ▼ as long as twenty-eight inches—on its head.

▲ Conspicuous because of the white around the base of their tails, Grant's gazelles, which have deep black S-curved horns, are always prepared for flight.

the dry savannah and high isolated plains. But it also needs water, and it wanders to find it. Thus we see herds roaming hundreds of miles in order to satisfy their thirst or hunger. The gazelle has a tiny cousin in the pygmy antelope, which is the size of a large hare or small dog, and its largest relative is the Cape eland, which weighs over a ton. The giraffe-antelope, or gerenuk, is slender, with an elongated neck; it stands on its hind legs in order to reach the succulent acacia leaves on which it feeds and which quench its thirst. When frightened, the impala, the black-heeled antelope, takes flight with an enormous bound that propels it up to nine feet in the air and a ground distance of thirty feet. Grant's gazelle is even faster, and only the cheetah, the fastest animal of all, can sometimes overtake it.

Antelope have one essential occupation: grazing. They eat endlessly, but when they graze in the open they are careful to post lookouts on nearby hills, or, lacking that, on termite nests, which may rise twenty feet high. Because danger is everywhere on the open plains and the herd may have to flee at a moment's notice, antelope swallow plants whole and later regurgitate, chew, and digest in peace when they are safely hidden in the trees.

Males and females, with the exception of old males, live together all year, except during the mating season, when couples leave the herd together, and at the time of birth, when the mother hides her young one, helpless and fragile as a newborn fawn, in the shelter of thickets, until it can follow the roaming of the herd.

◀ Impalas (*Aepyceros malampus*) are cautious, avoiding open spaces and remaining on the edges of light forests and along rivers. Only the male impala is crowned with lyre-shaped horns.

▲ The acacia provides much nourishment, and is the favorite food of the giraffe-antelope, or gerenuks (*Lithocranius walleri*).

◀ Gerenuks take refuge in a bush in the savannah. They leave it only to take rapid flight.

The Cheetah

The cheetah is often described in the language of auto racing: speed attained per second, standing start, miles per hour. With its long thin legs hard as steel, its arched body, its muscles tight as springs, it embodies esthetic perfection as well as efficiency: and it is self-confident, indifferent, insolent.

The cheetah is in some ways more doglike than catlike. Its claws are barely retractile. It doesn't like to climb trees. Unlike other cats, it hunts in broad daylight and rests at night. The cheetah's hunting technique bears no traces of fair play: it follows a herd for some distance, seemingly relaxed and indifferent to the temptations of this world. Then, suddenly, the cheetah pounces, fells the prey animal with a paw, and kills it with one crunching bite. The only fault of this admirable machine is the threatened prey's only hope: endurance in speed. The cheetah, a wonderful sprinter, cannot maintain its starting speed for long, and must give up after about 500 yards. But few animals last this distance at 75 miles an hour, and the odds are against the intended victim.

The different races of cheetah are found from India to the southern Sahara, across Persia and Mesopotamia. All are about the same size — two feet tall and four and a half feet long, with a tail of about three feet — and have a thin head with round eyes and ears. The Indians call it *chita*; thus our English name.

The ancient Egyptians and the Mongol and Indian princes used the cheetah for hunting. The Great Moghul of Delhi, according to the chronicles, would take as many as a thousand cheetahs hunting with him. They are used for the same purpose even today. Only an adult cheetah that has

The long, loose muscles and steel-like motor under the cheetah's spotted coat are the ideal structure of the hunting animal. The cheetah's favorite territories are the plains and desert.

Opposite:
Longing for freedom and a thirst for the old wandering life can be read in the eye of this cheetah (*Acinonyx jubatus*), whose species has been enslaved by hunters since antiquity.

▲
A peaceful scene: cheetah cubs cut their first capers under the mother's watchful eye. At this age they are no more formidable than large kittens.

lived in freedom can be trained. As a rule, training lasts six months from the time of capture, and the animal is never caged during training. When the animal is considered ready for service, its head is covered and it is brought thus blindfolded to the vicinity of a herd by cart. Once freed, the cheetah takes only a moment to orient itself, its old hunter instinct and technique coming to the fore again. When it has fixed its prey it bounds on it, slamming it to the ground with its front paws. It then waits for its share of the spoils—the blood that the human hunter spills into a wooden bowl by slitting the throat of the felled animal.

Like all other cats except the lion, the cheetah lives alone except during the short mating period; however, once mating has occurred the two partners part forever. The female brings up the young (usually two to five cubs), which are born trembling, furless, and blind, with the same care a domestic cat gives her kittens. She feeds them and tends them until they can live independently and alone—unless, of course, they are captured and turned into hunting partners for man or, worse yet, become show animals in a circus, trained to perform like dogs on a leash.

▲
◀ Learning to hunt: the mother teaches the young how to track, follow, bring down, and kill a prey animal. This stage of family life does not last for long. Solitude begins as soon as apprenticeship is over.

The Chameleon

▲
Frozen to its branch for hours, the chameleon watches an insect that will be its prey.

Reptiles appear slow, and the chameleon can look the slowest of all. There is a reason for its only temporary spurts of activity. Like the other members of its class, its heart does not separate oxygenated from deoxygenated blood, and its circulation is less efficient. It has extraordinary talents, however, that compensate for its methodical slowness.

The kaleidoscopic changes in color and design that this reptile undergoes are so striking and well known that the word "chameleon" has become a common metaphor. It is less well known, though, that the chameleon's color changes depend not only on its physical background, but equally on changes in light, temperature, or even the animal's "mood." Anger or fear can cause pigmented cells to shift, causing the chameleon to turn from light to dark, or from green to brown. It is often difficult to spot a chameleon hanging by its tail in a tree, its body flattened out in leaf shape. Twiglike extensions project from its nose like a horned and pointed helmet; colored occipital lobes and dorsal crests complete the disguise.

A master of camouflage, the chameleon

This photograph of an African species clearly shows the rotating eye capsule that gives all chameleons an extremely wide visual field. ▶

▲
Crowned with three horns that make it resemble a tiny dinosaur, Jackson's chameleon (*Chamaeleo jacksonii*) is more comical than frightening.

Fused with the tree that shelters it, *Chamaeleo bitaeniatus* is hard to spot. ▶

▲ Thanks to its skillful pincer-like hands, with three fingers on one side and an opposable two on the other, the chameleon has an excellent grip, made even more useful by long claws that curve inward.

Opposite:
The chameleon's long sticky tongue darts out rapidly, picks up a distant victim, and swiftly brings it to the mouth.

Anger, hunger, or fear can cause this desert chameleon to change from yellow to brown. These "emotional" color changes are not protective and can even risk exposing the chameleon. ▶

is also a champion marksman. Its weapon is its sticky lasso-like tongue, which is as long as the animal itself. The chameleon projects this tongue toward its prey—generally an insect, though sometimes, in the case of Madagascar chameleons, a mouse—with remarkable precision. It turns its eyes, which are enclosed in round capsules and can move independently of each other, in any direction—it can even see behind its head. Thus, like a crack shot, the chameleon can estimate precisely the distance between it and its next meal.

Most chameleons live in tropical Africa and in Madagascar. Diurnal creatures, they hunt for food only during the day and sleep at night. Even when they are in the middle of preparing the pit in which they will later lay their eggs, the oviparous females interrupt their hard labor—digging an eight-to-twelve-inch hole in the ground, filling it with dirt, then camouflaging it with plant debris—each evening at dusk to return to their tree for the night. The young, born after three to ten months' gestation, are capable of fending for themselves from birth. This is fortunate for them, since their parents immediately abandon them.

The female of a South African species of oviparous chameleons has no such need to exert herself in preparation for birth. She lays her eggs without even having to leave the bush. The newborn enter life covered with a sticky membrane that attaches to a branch. As the young chameleon begins to move about, the membrane tears. It then hangs from the tree by its legs and tail, finishes the task of freeing itself, climbs down, and goes forth to live its life.

While the male chameleon is a totally indifferent father, it has traces of a lover's temperament. The mating period is the only time when it becomes slightly animated. In order to discourage possible rivals, the male blows himself up, pricks up his crests, feints shamelessly, going so far as to actually fight if necessary. And to seduce his mate, he shows her his red belly.

The Giraffe

If the lion is king of the savannah, the giraffe may be considered the queen. The giraffe has no natural enemy; even the lion dares not attack it. If the lion were to pounce at its head, as large cats do, the giraffe would crush the lion against the nearest tree. If the lion were to jump between the giraffe's front paws, the giraffe would crack its skull with one kick. Because of its great height the giraffe acts as a lookout for other animals: antelope and zebra come to it for protection, and when one of their own natural enemies appears, the sentinels give the alarm by raising their tails. The giraffe is a pacific, meek animal; the Arabs called it "zurafa," or "the graceful one."

During the Quaternary period giraffe were found in India and throughout southern Asia, but today they exist only in the African plains and savannahs, where they can be observed posing, like stars, for

the cameras of the picture hunters. Their bodies can be as long as seven feet, and they stand eight to twelve feet tall. Their inordinately long necks have only seven vertebrae, like man's. They move in leaps of twelve to fifteen feet, gracefully using their necks for counterbalance.

Giraffe live in family groups, each male apparently attached to several females. These families gather in small herds of about thirty animals, and graze together on the leaves of fresh thickets. If the leaves are fresh and succulent, giraffe can do without water.

Every two years the female gives birth to one baby giraffe after a pregnancy of fourteen to fifteen months. At birth the baby weighs sixty-five pounds and is about

four and a half feet tall. An hour or so after birth it rises shakily on its long legs, and the mother, who had discreetly retired for the birth, returns to the herd with the little one. The baby is then handed over to a collective "nursery," which is watched over tenderly by the entire herd, though each of the young is the special responsibility of one particular "nursemaid."

Giraffe live for about twenty-five years, quiet, respected, pacific. At mating times the males engage in a strange combat, swinging their necks and knocking their heads together. But this is only a sham fight. One of the adversaries very soon gives up, and without a trace of a grudge he caresses the victor to demonstrate his peaceable feelings.

▲
Thanks to its brush-colored spotted coat, giraffes like soldiers, live in constant camouflage.

Opposite:
The giraffe (*Giraffa camelopardalis*) grasps tree leaves with its long blue tongue.

A baby giraffe attempts its first steps with the help of its mother. Only half a day old, it can already run, though at first it does not go far. ▶

▲
The giraffe in its domain, the endless spaces of the plains and savannahs.

Mock combat: two males confront each other over possession of the female. However, this is simply a show of dominance, and serious wounds are
◀ rarely inflicted.

The giraffe is always marvelously graceful, whether stretching its neck to survey its surroundings when in repose (*left*), or bending down with dignity (and straight knees) to drink.

Zebra, Hyenas, Vultures

The zebra is playful, enjoys motion, and betrays ▶ its lack of seriousness by jumping and capering. Zoologists have not been able to decide whether its striped coat serves as camouflage. Chapman zebras (*Equus burchelli chapmani*), which live in southern Africa (Zambia, eastern Bechuanaland, and the Transvaal), are yellowish, with black stripes and intermediary brown stripes. In the larger subspecies, which live in the extreme south, the contrast between the stripes is toned down.

The untamable plains zebra looks rather like a pajama-clad horse. But it has neither the horse's endurance nor the docility that makes the horse a friend to man. No one has ever succeeded in domesticating zebras.
▼

"Fast as a zebra" is a common expression, and the zebra is often called the "striped horse." This actually isn't quite fair to the horse, which is much faster than its striped cousin. However, like the horse, the zebra does enjoy galloping and running, rearing back whinnying and rolling joyfully in the dry grass of the plains. But unlike the horse, the zebra never allows itself to be tamed.

The great herds of zebra of the plains roam eastern and southern Africa, almost always in the company of antelope, giraffe, and ostriches. The complementary characteristics of the different species assures the security of the whole horde: all for one and one for all. Only an occasional imprudent animal leaves the group and risks being attacked by a lion. And even the lion fears the small iron-like hoof which the zebra uses to defend itself when it hasn't been able to flee. Hyenas are careful to avoid coming too close to zebra. But, in any case, of course much of their food is not alive, but dead.

Very little is known about the origins of these maligned animals that cackle like witches at night, emit a nauseating odor, and look rather like rangy wolves. Hyenas are thought to share common ancestors with the felines. There is no doubt, however, as to their usefulness. Guided by their unerring scent, they travel the brush or the savannah in quest of dead game or the leftovers of decaying food; they are responsible for the disposal of cadavers. They attack live animals only when they cannot find enough carcasses.

Hyenas live in couples or small groups. They prowl for food at night, and sleep during the day in caves or in holes that they dig, where they also shelter their young. The father and mother care for the young together during the nursing period.

◀ Childhood, even for hyenas, is a time of grace. With their large, round heads, these future sanitation workers still look innocently sweet.

▲ Solidarity is the great strength of the spotted hyenas, the largest in their family. They roam in hordes of about thirty. They are not content with carcasses and may attack grazing animals and even men. This is why they have been called "tiger-wolves."

The howling laughter of the hyena inspires terror. But one should be grateful for the clean-up job it performs and the scourges of disease that it ◀ prevents.

An elephant carcass doesn't impress the white-necked vulture (*Pseudogyps africanus*) of the African plains. In a few hours only a handful of bones will remain. ▶

The vulture shares the job of scavenging with the hyena. Man, who often views it as a bird of ill omen, generally finds the vulture as unattractive as the hyena. A flight of vultures over a caravan was seen as a sign of death by the nomads of old. There is little about the vulture to elicit warm feelings: neither its featherless ugliness, nor its predilection for the rotting food it so gluttonously devours.

There are many species of vultures. The white-necked vulture, or *Pseudogyps africanus*, is the grave-digger of the game-filled plains and savannahs of central and southern Africa. White-necked vultures, hyenas, and jackals, which have the same taste for cadavers, will often fight over remains. This species can be as long as three feet and have a wingspan of over six feet.

The Egyptian vulture, which is smaller (about two feet), collects the garbage, refuse, and excrement of all of southern Europe, Africa, and southern and western Asia. It nests in the immediate vicinity of man, in gardens, cemeteries, on temples, pagodas, or pyramids, with no respect at all for sacred places.

The father and mother sit on one or two eggs, and then feed the fledglings for a month, first with what they have stored in their gizzard, and then, after "weaning," with small prey: bats, frogs, or snakes.

Most vulture species live monogamously, though fidelity is not very strictly observed.

The Egyptian vulture (*Neophron percnopterus*) has dirty white feathers and is easily identifiable by the hairy collar that frames its face. It lives in the neighborhood of man. ▶

This peaceful behemoth charges only when it feels ▶ threatened. It then becomes a two-ton tank.

The Rhinoceros

Approximately sixty million years ago, rhinoceros already existed, though they were no larger than modern collies. Of the thirty-four recorded species, only four survive today. These weigh two tons and are nine to thirteen feet long. The African species (*Ceratotherium simum* and *Diceros bicornis*) live on the plains, avoiding the brush and forest. The Asiatic species—the armor-plated rhinoceros and the Javan and Sumatran rhinoceros—are limited to marshy forests, grasslands, and bamboo jungles. All have the same single need: proximity to water. They cannot live without their mud baths.

Rhinoceros are voracious eaters: they consume more than forty pounds of food a day and drink 90 to 110 quarts of water. Yet they are peaceful herbivores, almost phlegmatic. Only fear can transform them into furious monsters. Then they charge like wild bulls, overturning everything in their way, including men.

Rhinoceroses live in families, and when they change grounds they form a procession that follows a standard order: father, mother, and child (the child, which is born after an eighteen-month gestation period, is the size of a small pig at birth, and the mother nurses it for two years). They do not like to move, as they are closely attached to their territory. They have a strange habit that links them to their fellow rhinoceroses and to their territory: they have a common spot for defecation, where all the rhinoceros of the area gather for this rite. Even when rhinoceroses are fleeing they wait to reach this spot before relieving themselves.

The truly dramatic characteristic of the rhinoceros is its horn, which is an amazing pointed weapon that grows continuously from the corneous fibers of the epidermis and which grows back if it is ripped off. Rhinoceros horn used to be considered an aphrodisiac: it was traded for gold, and hunters amassed fortunes. In Asia today certain peoples still believe in its magical properties. The persistence of this belief is responsible for the extermination of the Indian rhinoceros, and the African rhinoceros owes its continued existence only to protective measures.

▲ Sleeping is the ideal pastime of the diurnal *Diceros bicornis.* Only at dusk, and with no excessive haste, does it go searching for food.

The rhinoceros loves to wallow in mud, which it could not do without. ▶

◀ The horn of the rhinoceros, reduced to powder and placed in a brew, is thought to increase sexual excitement. Rhinoceros almost perished as a species because of this belief. For them this horn is a weapon, though they will fight only a sexual rival.

▲ Pecking birds are rhinoceroses' willing helpers, living with them and ridding them of vermin, as well as warning them of approaching danger.

The profile of an eagle: a dagger-like beak, an intent gaze. *Aquila heliaca*, the true imperial eagle, has war written all over its face, though it rarely fulfills this appearance.

The Imperial Eagle

On good terms with man, here an eagle perches on a gloved hand.
▼

The imperial eagle, as well known as the stork, nests and broods either on the ground or in trees near man. It soars over the plains and marshy regions of the Mediterranean countries and as far as Mongolia. Ornithologists agree that the title "imperial" may have been bestowed a little hastily. Other species—notably the golden eagle, which dominates solitary areas such as mountain forests, sheer precipices, and desert plains—seem far more imperial than the imperial eagle, which in comparison seems small and mundane.

But when the imperial eagle rises with its wings open to their six-foot span, it seems, if not imperial, at least powerful and princely. But it is lacking in dignity. Its victims are ignoble (rats, mice, hamsters, rabbits), and it hunts them without risk of danger. It often even acts as a garbage disposal service to thankful villagers. In southern Europe it often uses telephone posts as oberservation points. From there it watches for marmots and susliks, which are a kind of ground squirrel that it preys on.

The female lays two or three eggs in the eyrie, which is a vast platform sometimes built on the branch of a large abandoned tree. Nine or ten months after hatching, the baby eagles fly to sunny central Africa or southeast Asia, as proud as if they were wealthy tourists leaving for a safari.

The Elephant

Elephants are part of history: Hannibal's elephants were sent across Spain and the Alps to conquer Rome. And there are countless stories, old and new, about the fortunes and misfortunes of this great beast. Elephants appear in Oriental legends, which are full of maharajahs poised on fabulous white elephants, as well as in the Western picture books about Dumbo, the flying elephant, and the lovable Babar. But elephants had to await the indignation of Romain Gary's *Roots of Heaven* before the world could be moved by the fate of these pachyderms which are threatened with extinction by hunting, both for trade and for glory.

The elephant is a highly paradoxical animal. Its ancestors, the as yet trunkless first proboscidians, lived in North Africa, and their prehistoric descendants spread all over Africa, Asia, Europe, and North America. They were no larger than tapirs. Today their Indian descendants weigh four tons and their African cousins, the largest of all living mammals, sometimes more than six tons. The elephant holds the record for absolute force, but it is also endowed with a great memory and intelligence—trained elephants can often understand human language, and not only obey their trainer's orders but understand their explanations as well. It appears that elephants also have "emotion," and thus may behave irrationally, just as does man. The elephant's cumbersome trunk is one of the most highly perfected organs of the animal kingdom. Formed by the elongation of the nose and the extension of the upper lip, the musculature of its sensitive tip permits a grasping action much like

Elephants, the last of the great pachyderm species, are under strict protection. The creation of great reserves should prevent the extinction of the so-called bush elephant (*Loxodonta africana*). But in unsupervised freedom, the great herds are becoming rare. ▶

The African elephant, the largest and heaviest of ▶ all the mammals, lives in savannahs and forests from the Sudan to the national parks of South Africa. Fear or solitude in old age can make it aggressive.

that of fingers and opposable thumb. The elephant uses its trunk to bring down the branches on which it feeds, to reach the high-growing fruit, or to uproot young trees in order to eat their leaves more conveniently. The elephant breathes and trumpets through its trunk, and uses it for scenting, touching, clasping, and showering. Its trunk is essential for survival.

This mammoth is a nomad, always searching for greener pastures. It likes company and travels in herds of thirty to a hundred elephants. It is capable of traveling four to five miles in a day, at times climbing mountains as high as twelve thousand feet, unstopped by rivers, swamps, or lakes. It has the wisdom to rest at night and nap during the hottest hours of the day. Elephants sleep on their sides in damp thickets, digging a nest for their heads in the bushes.

Though one of the more imposing males dominates the herd, in the great migrations it is often a wise and experienced female that actually leads the single file. In case of danger, however, this female yields her place to the male. The female is an attentive mother with infinite patience. She devotes two years of her life to each baby, not counting the gestation period of twenty to twenty-three months. During the time of infant care she pays no attention at all to the opposite sex, which explains the four-year period between each birth. The newborn elephant is thirty-two to forty inches long and weighs over two hundred pounds. It nurses until its second year, not with its trunk but using its lips to suckle on one of the two teats between the mother's front legs.

As long as the baby elephant cannot

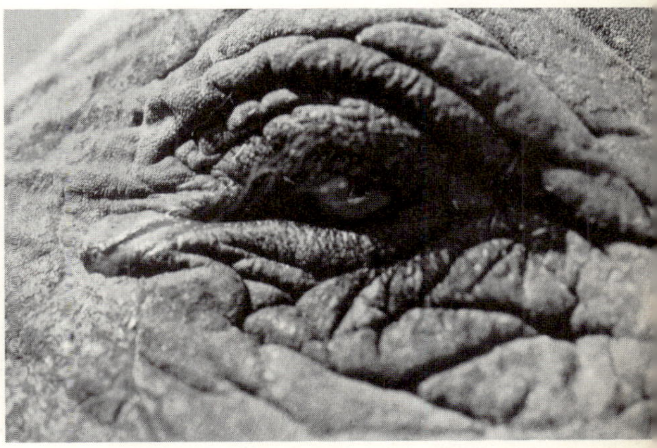

▲
The elephant's tiny eye, surrounded by a relief of ridged and wrinkled skin, seems vastly disproportionate to the animal's bulk. But the elephant can see well enough.

◀ The elephant pulls back its huge ears when it is hot, their large area providing cooling surfaces. It uses its tusks for digging, as leverage, and for leaning. Its trunk has universal uses: grasping food, drinking, showering, hitting, blowing sand, and loud trumpeting.

Though always on the move, elephants do know the joys of resting. They stop at watering points where they refresh themselves, bathe, and spray their playful babies.

walk, the herd waits for it. On the day of departure, it hangs on to the mother's tail with its trunk and follows behind, never leaving her. In case of danger, even if both mother and child must run, it is quickly hidden under her body. The young elephant reaches sexual maturity only at fifteen. This extended childhood is proportional to the elephant's longevity: it lives fifty, sixty, sometimes even a hundred years.

The elephant has one main occupation in life: feeding itself. It consumes 165 pounds of plants and 55 gallons of water a day. Its terrifying tusks are sometimes as long as nine to eleven feet in males and six to eight feet in females, and weigh thirty-three to forty-four pounds. At the turn of the century elephants with tusks weighing 130 to 155 pounds were still found. But tusks are used more often for digging up tubers than for combat. They are rarely needed in fighting: even the lion seldom attacks any but the young, and that rarely, since it would have to face the mother's devastating vengeance.

▲ Baby elephants often get tired walking, so they hang on to their mothers. As she moves forward, the mother cleverly sprays sand to drive away insects.

The Kangaroo

The kangaroo, emblem of Australia, the very image of jumping, and the symbol of tender maternal care, is a living fossil. It belongs to one of the oldest mammal groups, the marsupials, which were already present in the Cretaceous, some eighty million years ago. Its ancestors, which were small opossums, were scattered all over the world. But the kangaroo evolved only in Australia, that island continent, where more highly evolved mammals appeared only in more recent times and where the kangaroo is not the only trace of prehistory.

Apart from its jump, the kangaroo is best known for the pouch on the mother's belly in which the young is kept. After a gestation period of thirty to forty days, about two hours before dropping the embryo (which is only one inch long and $^{1}/_{30,000}$ the weight of the mother—in man the ratio is 1 to 20), the mother opens her pouch with her front paws and licks it meticulously. Sitting on the base of her tail, she gives birth to a tiny, red, furless, blind, and deaf baby. The tiny blob has only its front limbs developed. It climbs from the birth opening into the pouch where it attaches itself to one of her four teats, which will grow in the shape of a tuber and will continuously nourish the baby with milk. At three months, the baby sticks out its head for the first time, and at 111 days its fur begins to appear. At 250 days, weighing no more than five to nine pounds, it undergoes a second birth, emerging from the mother's pouch into the outside world.

The baby kangaroo is never an only child. Observations have been made of great gray kangaroos showing that while 75 percent of the females are raising one young in their pouch, 20 percent of these are at the same time nursing another young, and 60 to 70 percent also carry an embryo in their uterus. The embryo is patient: it develops only to a certain stage and waits until the older sibling has vacated the mother's pouch before continuing to grow and then moving in. This is why it is said that the female kangaroo, with her unfinished baby ready and waiting to be born and another one finishing development in the pouch, is more of a mother than a mate.

Kangaroos live in family bands, led by a strong male who is fairly old and conscious of his prerogatives. When a younger rival covets his harem, a fierce duel ensues, with mighty blows struck by the powerful hind legs. Sometimes, though rarely, kangaroos attack men. Ordinarily they are satisfied to be harmless and fearful herbivores, feeding on leafage, young shoots, bark, grass, or plants. They are tied to their territory, which they will defend against other bands, and leave it only for seasonal migrations, and only at night.

Great gray kangaroos can be two to three yards long, including their tail, and they can move at forty-five miles an hour in upward leaps of six to twelve yards. When they jump, their hind legs act as a springboard and their taut, thick tail as a balancing pole. At rest, the tail is a support to lean on.

Great gray kangaroos used to inhabit all the wide open spaces of the Australian bush. But when the Europeans arrived, they turned the bush into grazing ground for their animal stock. And as the Europeans developed a taste for the apparently highly edible kangaroo flesh, they hunted them, diminishing their numbers so drastically that in order to save the species kangaroos were even raised in European zoos and then sent back to their native land. Today only small bands of great gray kangaroos survive in the bush in the east, southeast, and south of Australia. These kangaroos live four to eight years. Those that live in zoos can reach seventeen or eighteen years of age. This is perhaps at least one compensation for their lost freedom.

The red kangaroo (*Macropus rufus*), which is five and a half feet tall, holds the record for jumping: a ground distance of thirty-six feet, nine feet in the air.
▼

Opposite, above:
The great gray kangaroo (*Macropus major*) reaches a height of four and a half feet. Of the sixty-six varieties of kangaroos living in Australia today, this is the best known.

▲
The female kangaroo is usually a triple mother. She is still nursing a fairly independent baby while the second child is completing development in her pouch and a third baby is waiting to be born.

2. Deserts and Semi-Arid Lands

The Ostrich

The ostrich, the giant among birds, has a head poised like a beacon nine feet above the ground. As it is too heavy to fly (it weighs about 220 pounds), it must run. On its tall muscular legs, which are as wiry and tireless as those of a horse, it can reach a speed of twenty-eight to thirty-two miles an hour. As it flees in strides of three or four yards, its short wings act as a balancing pole. In order to scan its surroundings, it keeps its long neck upraised as it moves along. Nothing escapes its sharp vision. This is why zebra and antelope follow behind: with ostriches as their lookout, they can graze in peace.

These strange birds live in the plains, savannahs, and the edges of the African deserts south of the Sahara. They are adapted perfectly to life in the great spaces, where they roam nonchalantly in herds of about twenty in search of food: shoots, fruit, grains, small animals. Like all grain-eating birds, they also swallow pebbles, which facilitate the pulverization of the foods in the stomach. They get their water from water holes or, in times of great thirst, from succulent plants.

The ostrich is polygamous. Each male rules over a harem, though he must work for it: it is he that digs in the sand and

▲
Marching order: when they are searching for food, ostriches (*Struthio camelus*) move in groups, roaming the plains and the fringes of the African deserts.

▲
At rest: the giant bird is hunched on its long muscular legs. While all other birds have three or four toes, the ostrich has only two, with short, powerful nails.

On the lookout: nothing escapes the ostrich's large brown eyes. Long-lashed and marble-like, they observe the world from a yard up. Wherever ostriches are standing watch, antelope and zebra
◀ can feel secure.

rounds the nest in the shape of a trough with his heavy body. Up to five females then deposit their eggs in this nest, and in two days each female lays about twenty eggs weighing two to three pounds—twenty-five to thirty times the weight of a chicken egg. The female, quite inconspicuous in her graying brown plumage, broods during the day. The male, whose brighter plumage requires the cover of darkness, takes over at night. After forty to forty-two days, the baby ostriches break through their shells and are immediately able to stand up. The father then takes over their upbringing. Sometimes several family heads create a "nursery" where young are raised collectively.

The females are capable of reproduction around age three, the males at four or five. Before taking a mate, the male goes through a ballet of seduction. He circles the female, spreading his snowy white wings for her admiration, struts up and down, crouches on his legs, swings his body back and forth, and arches or puffs up his neck, all the while making growling but tender noises. The dance isn't restricted to mating, and sometimes when an entire herd is carried away by exuberance it may execute, in the middle of a plain or a dune, ballets worthy of Maurice Béjart. The dance of the ostrich is one of the classics of African folklore.

Just born, the fluffy and shaggy baby ostrich can ▶ already stand up.

The eggs, guarded jealously, are usually those of several females. Brooding lasts about six weeks, during which the eggs, which weigh two to three pounds, change from yellow to ivory. The duller-colored females brood during the day, making themselves as squat and flattened out as possible. The multicolored and shimmering male takes over at night. When attacked, the ostrich flees only under the direst circumstances, often pretending to be sick in order to divert the enemy from the nest.

The Gecko

The gecko's adhesive fingers are unique in the animal kingdom, and it is only recently that their secret has been discovered. Their webbed fingers have thin sheets of skin of all shapes and dimensions on their undersides that can stretch in any direction. These are covered with several million very fine bumps that solidly grip the most difficult surfaces. In addition, the skin is wrinkled and stretches and folds back, thus creating a vacuum between the animal's paws and the supporting surface. This suction-cup system allows the gecko to stand anywhere. Attaching and letting go take place by a simple separation of the wrinkles, and occur so rapidly that they cannot be seen in detail. Most geckos also have strong and retractile claws, which they use as clamps. Finally, some desert species have regular snowshoes: their toes are connected by large webs with fine scales. These allow them to run at full speed on the surface of the sand without sinking in.

The gecko's eye is extremely accurate. Some varieties have a pupil divided into four openings, which simultaneously send four images to the retina. The nervous system coordinates them into a single image with great depth of field.

The gecko's tail is equally startling. Sometimes it is as flat as a leaf, and serves as a support as well as a crutch in running and climbing. In other species, it grows in

There are no obstacles for the wall gecko. This shy inhabitant of Mediterranean countries, which is yellow, grayish-brown or dull black, can get a grip anywhere thanks to the adhesive scales on its fingers and the clamplike claws on its third and ◀ fourth toes.

Geckos have no human enemies; they are welcome in homes because they kill insects. Most geckos live in India and Southeast Asia, but some species occur in other tropical regions. Below is the largest of the geckos, the tokay. It can be seen running about everywhere, even on windowpanes, in its coat of light gray, bluish, or white scales, sprinkled with red dots. Like a cat, it catches mice and, like a dog, it sometimes barks.

The gecko's foot is an all-purpose tool, unique in the animal kingdom. Its toes, which have cushions of transverse sheets of skin covered with very fine bumps, can grip the most difficult surface. On an absolutely smooth surface, it folds and unfolds the skin between the toes, and its whole body adheres by the suction cups thus formed.

The gecko cries, "Tokay, tokay" as it runs through the house. Snarling and barking, it can defend itself well and does not hesitate to take on domestic animals.

the shape of a turnip, and is used to store fat. The attacked gecko, like many lizards, abandons its tail to its enemy. The adversary will hesitate with fascination before the tail trembling savagely before its eye, giving the amputated gecko time to run away. A new tail will soon grow. And if the entire tail hasn't been lost, the old tail will join with the new one, making it a two-tailed gecko.

Geckos are found in all hot climates, especially in the deserts and dunes of Africa and in India and Australia. Most are barely six inches long and have a heavy, almost flat body, a large head and short legs. They feed on flies, gnats, spiders, cockroaches, caterpillars, and—for they can bite fiercely—mice. All species, except for a few in New Zealand, generally lay two eggs, which the female leaves under the bark of a tree, under a rock, or in the crevice of a wall. Geckos have even been known to lay their eggs behind window shutters or under a carpet. The eggs hatch after several months.

This enigmatic animal was long misunderstood: it was thought to be poisonous. Today we know that it is not only harmless, but even useful. Asians believe that it brings good luck. The largest geckos—ten to twelve inches—often live near man.

The gecko's tail is a weapon of diversion: seized, it detaches as the animal flees. In the deserts and dunes, where geckos are numerous, these strange little lizards dig walls and dens.

▲
The gecko's eye is a precise instrument; it defies the night. Most gecko species are nocturnal, and the pupils of their eyes, like those of a cat, thin to a slit. The wall gecko can be recognized by its eye, which has its pupil broken up into four sections that transmit simultaneous images to the retina. These images are coordinated by a nervous system that gives the wall gecko perfect vision.

A time of labor: during the last decade, the trading fleet of the men of the desert, moving in the burning heat of an arid land, was made up of four million dromedaries (*Camelus dromedarius*) and two million Bactrian camels (*Camelus bactrianus*).

Camels

A time of freedom from care: until the age of four, young camels trot burdenless beside the caravan. Later, they will serve man and carry their load for twenty-five to thirty years.

The camel was already serving man three thousand years before Christ in the high plains and mountains of central Asia. A thousand years later, in the desert expanses of Arabia, the dromedary became the companion of the Bedouin, whom it aided in conquering Africa. Today, despite the competition of the truck, camels and dromedaries continue to be the main ships of the desert. Local trade in the Sahara is still conducted by caravan. In the last decade, four million dromedaries and two million camels, both as mounts and as beasts of burden, constituted the traveling "fleet" of the men of the desert.

One couldn't wish for a stronger, more sober, or more enduring servant. As a mount, it can trot sixteen straight hours and cover about ninety miles in a day, carrying on long voyages a load of 330 pounds and up to 880 pounds for short distances. Its body is perfectly suited for the aridity of its environment. Though it loves fresh grass, it is usually content with hard, dry, spiny bushes which it gulps down into its first stomach and later peacefully chews. When there is no food, it first uses its own bodily reserves for nourishment and water, and then the hump of fat. It can go completely without water for four or five days, and lose up to one-fourth of its weight from thirst without collapsing; camels can be observed at oases gulping down as much as thirty-five gallons of water in ten minutes. In a sandstorm, the camel can almost completely seal its nostrils, and its eyes are protected by a double row of long lashes and when necessary a flood of tears. Its legs are protected from the desert's pointed pebbles by thick calluses. Its skin is thicker at the knees, elbows, and chest, the areas of its body that act as a cushion when the camel lies down on the hard ground.

The camel's family tree dates back to the Tertiary period, fifty or sixty million years ago. Its ancestors were no larger than sheep and inhabited North America, where their descendants became extinct during the Ice Age, but not before migrating to Asia and South America. The dromedary, which inhabits North Africa and western Asia, is about eight feet long and seven and a half feet tall at the withers. It has only one hump. The Bactrian camel has two humps and is about four inches longer than the dromedary. It lives in

central Asia from Turkestan to Mongolia. Domesticated camels live in conditions established by man. Wild camels gather in herds led by the strongest stallion. The mating period does not increase sociability: it cuts the male's appetite, makes him grouchy and complaining. When males confront each other for the possession of a female, they spit in each other's eyes and emit a repugnant odor from two special glands in the nape of their neck. They make a noise by squeezing an air-filled noise bag deep in their mouth, try to wound their adversary with their long front teeth, or knock it down and beat it with their front knees.

After a gestation period of twelve to thirteen months, the female leaves the herd to give birth to a furry baby camel. The baby is forty inches long at birth, and can soon stand up and run with its mother. In freedom, baby camels nurse for several years, but Bedouins wean the young camels by placing a sharp point in the animals' nose. This point is painful to the mother's teat, and she then turns the baby away.

For centuries this scene of long caravans winding through the Sahara has remained unchanged. Despite competition from the truck, the camel is still the natives' ship of the desert. ▶

The baby camel is almost forty inches long at birth. Soon it will run beside its mother in the pasture and then join her on the long caravan routes.
▼

▲ The protection of a double row of long eyelashes, which come together when necessary with a curtain of tears, and nostrils that can almost be sealed, allow the camel to face the terrible sandstorms of the Sahara.

Cushions of callused skin line the camel's knees, ▶ elbows, and chest, the areas on which the animal rests its weight when it lies down.

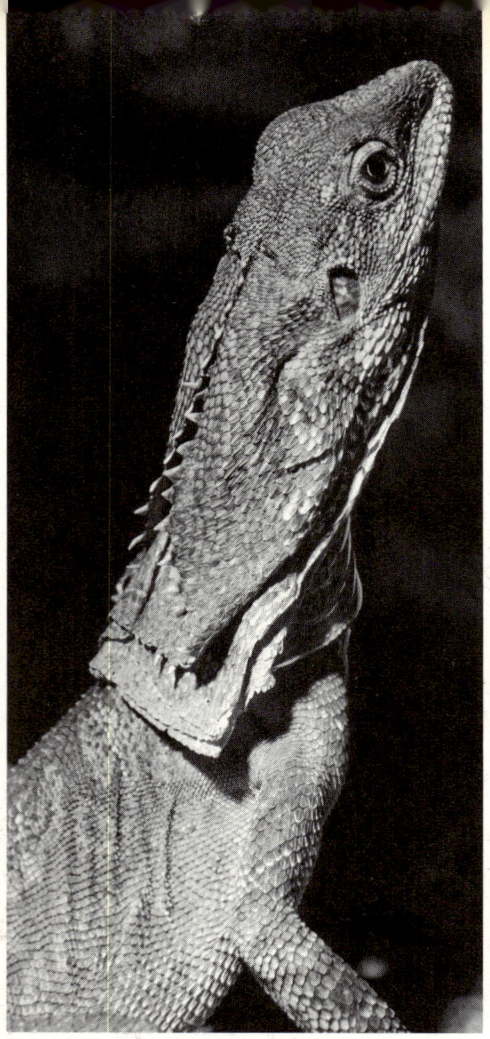

The Frilled Lizard

▲
The frilled lizard (*Chlamydosaurus kingii*) wears an armor of brown scales dotted with yellow, red, and black. Its collar, eight inches in diameter, can spread out into a bright red, blue, and brown ruff. The lizard uses this to frighten its adversaries and for display.

When fleeing or attacking, the frilled lizard looks rather like a hurried businessman, standing as it does on its hind legs, with its tail acting as a balancing-pole.

It is about forty inches long, has yellow, red, and black-speckled scales, and belongs to the agamid lizard family. While almost all its relatives inhabit the plains and deserts of Asia and Africa, it lives in Australia and New Zealand.

When threatened by an enemy or rival, the frilled lizard uses a whole gamut of means of intimidation provided by nature. First it charges, standing up; then, quick as lightning, it crouches, violently raising its front legs high; it opens its mouth, exposing its toothy jaw. If this isn't enough for even a psychological victory, it spreads its great collar of fine red, orange, blue, and brown mosaic-like scales in all directions. This is its second warning. The third time around, furious, it comes into direct contact and bites viciously, simultaneously using its tail as a whip.

Relatives of the frilled lizard, generally half as large, are each unusual in its own way. The bearded lizard's cheek-flaps swell and become an enormous spine-filled beard. The spiny-tailed lizard *Uromastix* has a tail covered with spines that it holds in its mouth as a protective device when it flees into rock crevices. And the flying "dragon" spreads out its large folds of skin, supported by long ribs, to glide from tree to tree.

When angered, these lizards can be rather frightening. However, this is more appearance than reality, for in fact they are harmless creatures that feed on insects, plants, and fruit.

◀ To flee, the frilled lizard stands up on its hind legs and runs like a hurried man, using its tail as a balancing-pole.

▲
To frighten, the angry male opens wide its toothy mouth. But when the adversary is not dissuaded by these threats the frilled lizard flees.

3. Tropical and Subtropical Forests

The Tiger

According to an old Indian proverb, "Wherever there is fear of the tiger, it is night." It is often night from Malaysia to China, from Java to Sumatra, from India to Manchuria, where tigers are to be found. The fierce tiger, the crafty tiger, the terrifying tiger, all are part of the demonology. The presence of this great hunter beast is revealed only by a thin track in the tall grass or an almost imperceptible trembling in the reeds. The tiger never shows itself openly. By the time it leaps, it is too late to flee. In just a few strides and a flashing outstretch it is upon its long-observed prey. Its paws hit so violently that powerful ungulates, such as wild pigs or antelope, roll to the ground. One crunch of the jaw on the animal's throat ends the brief drama, and the tiger begins its feast. But, always cautious, it first drags its victim into cover. If it is very hungry, the tiger can devour forty-five pounds of meat. Then it quenches its thirst lengthily and returns to the remains of its meal, on which it feeds for several days, until the last morsel, regardless of the state of decomposition. It will strike again only when its food supply is gone, satisfying itself in the meantime with mice, turtles, lizards, frogs, or even fish as hors d'œuvres.

The tiger's eyesight is poor—something unusual for felines. It is guided by ultrasensitive hearing and a sharp sense of touch. It hunts at night as well as in the day, lying in ambush alone, and remaining as far as possible from other tigers. When another tiger intrudes upon its favorite observation spot, it is chased away with fury. Though the tiger enjoys solitude, it enjoys special pleasures, too, such as bathing. Even in the coldest regions it swims and can cross wide rivers and inlets.

The tiger used to spread terror through all the forests of Asia, from the Caucasus to Korea, from southern Siberia to the islands of Java and Bali. Today, the best known race is the Bengal tiger, found only

▲ With one flashing leap, the tiger is upon its prey. Its long supple body is ideally suited for running and leaping.

The growling, excited tiger wears a demoniacal expression. But its hunting skill depends on the discretion of its furtive approach. ▶

in India. The Chinese tiger lives near the great rivers of central and southern China. The tigers that lived in northern Afghanistan and Persia seem to have been exterminated. The Siberian tiger, or Manchurian tiger, which is the largest and most beautiful and is strictly protected, inhabits the area north of the Amur River between Siberia and Manchuria. In Bali, there are only about twenty tigers—if indeed these have survived. The tigers of Java and Sumatra are under grave threat of extinction.

The diverse varieties of tiger have striking differences in appearance and behavior. The continental tigers are six and a half to seven feet long, while the island tigers are less than six feet long, with twenty-eight-inch tails. Those in tropical regions have short hair, while their Siberian relatives, which can weigh up to 550 pounds, are protected from the snow by long, thick fur padded with a wool fleece. In the south, mating depends solely upon the animal's choice, while in the north it depends on the seasons.

The male, always imperious, calls the female with great hoarse cries. But no feelings seem to be involved, and the encounter is always brief. After about a hundred days a litter of two to five clumsy little tigers is born. For a year the mother trains them to track and hunt. After this they lose all family feelings, and become solitary brigands like the adults. They are sexually mature at age three, and live

The tiger's striped coat is hard to see in sunny tall grass or reeds. The cat sneaks up to its prey, watches it carefully, and then attacks. ▶

Each mother very carefully teaches her young how ▶ to kill during their first year. This is the only time when tigers hunt together and share their gains. The young soon become solitary brigands like their parents.

This remarkable swimmer is bathing with obvious pleasure. Even the Siberian tiger doesn't hesitate to swim in the icy waters of its habitat. ▼

about a quarter of a century—unless they perish prematurely at the hand of man.

While lion hunting was the royal pastime of the pharaohs and rulers of Babylonia, tiger hunting was long the spectacular entertainment of the maharajahs, accompanied by hundreds of elephants and thousands of servants. "It was a dazzling and extraordinary spectacle," writes one chronicler. "A whole village of tents would house the potentate, his guests, and innumerable servants in the middle of the jungle. Poised on bejeweled elephants, the hunters would take up positions along the path which the tiger was expected to

use to flee, while legions of beaters on foot would hem in the forest. Poised high on his pachyderm, the hunter would be out of the tiger's reach. Moreover, the tiger, accustomed to seeing elephants which it occasionally passed in the forest, would approach without much distrust. All there remained to do was shoot."

In India today, the elephant has been replaced by the tree and the beaters by bait. This is hunting to kill: the hunter hoists himself into a tree near an animal killed by the tiger or near live tied-up bait such as sheep or buffalo. Here he sits, where the tiger, which cannot climb, cannot reach him. Then follows the pattern of all wild game hunting: waiting, accurate aim, and fast shooting. Searching the jungle for a wounded tiger is not very exciting, but following the trail of a known man-eater can be dangerous. Old tigers may reach the point where, their bodies wounded or their senses dulled, they can no longer hunt wild game. Then they become surprisingly bold; they may even go into villages in broad daylight and try to open doors and enter homes. A tiger can terrify an entire region this way. In Nepal, there is still talk of a tiger that killed over two hundred people.

Gorillas and Chimpanzees

The newborn was lifeless. Grasping its tongue between her lips, the mother revived him by mouth-to-mouth resuscitation. This could be a story about people, a human-interest item in the news. But it is an *animal* story, reported by zoologists specializing in the study of chimpanzees. Anthropoid apes are indeed the most intelligent of animals, and their morphology and behavior are the closest to man's. They are capable of foresight, and can establish a relationship between cause and effect. Their physical attitudes, gestures, and mimicry can express almost the entire emotional gamut: love or indifference, joy or grief, fear or defiance. They waste away when alone, as they enjoy life only in society. They are capable of making and using tools, without a model to follow. They have mastered the elementary principles of construction and build platforms of carefully interlaced branches in the trees, where the females stay with the young.

Gorillas and chimpanzees live in family bands made up of one or several couples and their offspring. With gorillas, this totals about twenty animals, including a strong male (the leader), four to six females, and young of various ages; chimpanzee groups are usually about ten, with more females than males. Gorillas do not reach maturity until the age of six; chimpanzees not until eight. Mating is not confined to any particular season: sexual activity is constant all year. After a period of gestation of seven to nine months, the female gives birth to a baby—twins or triplets are very unusual—which she immediately takes into her arms, as a woman does with her child; the newborn clings to the mother's breast near the source of milk. Frail and bare, he needs warmth and care. His mother lavishes affection on him, gazing lovingly at him, rocking and cajoling him, cleaning him with a kind of grave tenderness. He remains gripping her fleece for a month, even when she runs and jumps. Then, cautiously, holding him by the hand, she teaches him to walk. The father always watches over their play and their rest.

Gorillas are as timid as they are imposing, with their height of six and a half feet, their hundred-inch girth, their muscular 450 to 650 pounds. They live by themselves in the virgin forest of west Africa and in the Congo, between the Cameroons and Katanga. Their fleece, often black, can be as thick as four inches. They have little hair on the chest and abdomen,

Imposing, shy, and peace-loving, the gorilla (*Gorilla*), largest of the anthropoids, lives hidden in the virgin forest of Africa.

The face of the chimpanzee (*Pan*) can express seriousness, perplexity, decision, concern, humor, and parental love.

and the palm of the hand and the sole of the foot are bare. They use all four legs in running, leaning on the entire sole of the foot and then projecting themselves forward on their toes. In spite of their bulk, they move with agility through the trees of the forest as they search for food: leaves, fruit when available, birds' eggs, insects or baby birds, and especially bamboo shoots, their favorite food. It is always the strongest male who leads the tribe. He keeps watch on the ground while the clan rests in the trees, and he provides the rear guard in case of flight. If forced to attack to defend the group, he faces the enemy, rises to his full height, grinds his teeth, growls, and beats his chest fiercely. If this show fails, he rushes at his adversary, biting him furiously and bludgeoning him with his horny paws, which are powerful enough to shatter bones.

Chimpanzees, since they are more common and also bolder than gorillas, are more easily and conveniently observed. They live everywhere in the virgin forests of west and central Africa, from Gambia to Uganda, and in the northern part of the Congo. The males measure approximately 5 feet 8 inches and weigh 130 to 175 pounds. The females are smaller.

▲
Chimpanzees can use rudimentary tools and weapons. Dutch researchers observed a small group of these anthropoids attacking a stuffed predator: trap: first they menaced it, then they threw stones, then armed themselves with clubs in order to kill the cat.

Both sexes have black, or sometimes light beige or brown, fur.

To run on all fours, the chimpanzee rests his hands on the ground and places his feet outside them. But he is especially comfortable climbing in the treetops, swinging and springing from branch to branch in a dazzling aerial act. The liveliest of the anthropoids, chimpanzees are always bustling about, cheeping, gesticulating. They feed on fruit, bulbs, young shoots, tubers, insects, small birds, mice, and termites, for which they patiently "fish," introducing a blade of grass into the termite nest and quickly pulling it out as soon as the termite "bites."

Among themselves, chimpanzees are rough and not very cordial, often threatening each other, hands on hips, like hooligans. But they sometimes greet each other, as if it were the most natural thing in the world, by shaking hands. Furthermore, they can speak—or almost. Scientific studies have demonstrated that they can modulate a repertoire of about thirty sounds with their lips in communicating with one another. The era of discoveries about these disturbing and fascinating cousins of man is not yet over.

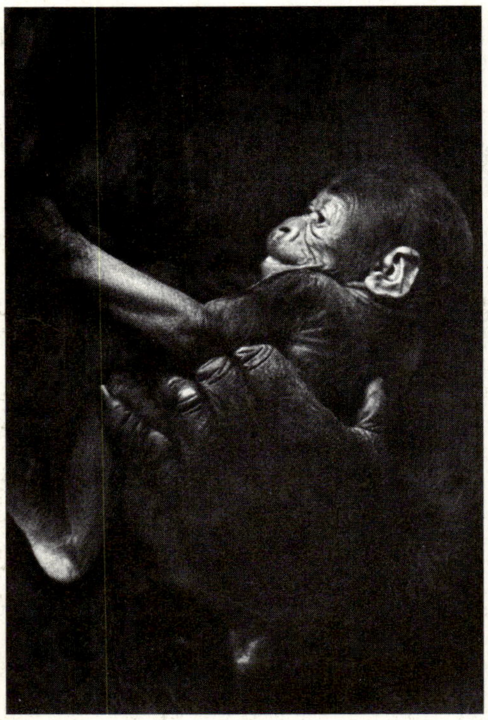

Like a woman caring for her child, the mother watches tenderly over her baby gorilla. *Right:* One of the rare photographic documentations of these animals in a free state shows the bed of branches high in the trees where the females sleep at night with their young.

The Leopard

The most beautiful, lithe, crafty, and determined of the felines are the leopard and its dark twin, the black panther. The leopard hunts at night, all its senses alert; its ears pick up every sound. Once beneath its iron and velvet paws, no prey escapes, and no shadow eludes its topaz eyes. It patiently waits for the moment to strike. It hunts monkeys, its favorite prey, through the branches in long, silent strides. It climb trees with the greatest of ease and pounces on antelope, jackals, rodents, bovines, and even buffalo and wild pigs. It roams through villages, entering stables and attacking livestock and dogs. It can be frightened only by fires and torches.

It kills efficiently, with one bite of its sharp-edged fangs that rip and tear. It then drags its victim into a tree, drinks its lukewarm blood, devours the entrails, and then finally consumes the flesh to the last shred on the last bone. It gulps down as much food as it possibly can, and if there are leftovers hides them between two branches for its next meal. As long as the leopard has food in the tree it will not hunt, but it will savagely defend its hoard against other animals and men. Natives know this, and will avert their eyes when passing the leopard's tree: one glance could arouse the animal's suspicion and prompt an attack.

Like most cats, the leopard hunts and

Comfortable in his tree top, the leopard (*Panthera pardus*) eats and sleeps there during the day; at night it hunts monkeys, its favorite prey, through the branches. The leopard is the handsomest, most graceful, and cleverest of the felines, and one of the most daring and dangerous. But it never kills more than it can eat.

With incredible strength and determination, this perfect cat hoists its prey—a gazelle, antelope, or monkey—often larger and heavier than the leopard itself, to the fork of a tree, where it wedges it in and then eats it.

lives alone. Only mating interrupts this solitude, and the male fights noisily with his rivals. After a short mating period, the male and female separate. Some ninety days later, two to five small blind leopards are born, with coats that are already beautiful. There may be a black panther in the litter; simply the result of a hereditary variation, this cub has the same spotted coat under her shiny black fur. This phenomenon is especially common in India.

The terror of men, the decimator of whole herds of animals, the beast of legend, the leopard lives throughout Africa except for the true deserts, in Sumatra, Java, Ceylon, southeastern China, India, Persia, all of Asia Minor, and as far as the Caucasus. It lives in dry and open areas, sometimes in virgin forests, and sometimes also in mountain forests. Hemingway even found a carcass on the snowy slopes of Kilimanjaro, and pondered what the dead leopard could have been seeking at so high an altitude.

The leopard has always been hunted for its sumptuous coat. The balance of nature is a fragile thing indeed: in places where the leopards were being killed off, baboons and wild pigs that devastate farmlands multiplied at such a fast rate that the natives, unable to farm their fields, had to emigrate. To save these lands, leopards are now strictly protected.

◀ Neither animal nor man dares dispute the remains of a leopard's victim — here an impala — hanging from the branches. The leopard will attack whomever it suspects of wanting to steal its prey, even by a single unwise glance. It makes no exceptions for man.

Pythons

One of the high points of most documentary films on animal life in the jungle is a duel to the death between a wild animal and a python. The snake winds itself around its adversary, which fights back desperately. Unrelenting, the coils of the snake tighten around the prey, which is sometimes as much as three times the size of the python, crushing and strangling it. Then, very slowly, the victim disappears, head first, into the horrible toothy mouth of the snake. This familiar and spectacular scene, which needs no exaggeration for effect, inevitably leaves the fascinated viewer horrified—and perplexed. How can the python, a nonvenomous snake, accomplish this?

The answer is anatomy. The snake's jaws and palate are connected only by elastic ligaments, and thus the mouth can expand to accommodate so large a prey. The two halves of the lower jaw are arranged so that they can move forward and backward independently. The curved teeth take hold of the victim and force it into the gullet, inch by inch. Throughout this operation, which can last hours if the victim is large, the opening of the trachea is pushed far forward. The prey—with skin, claws, bones, fur, and all—reaches the digestive tube, which is as long as the snake itself. Powerful juices are then secreted that digest everything digestible.

Pythons live in forests and humid tropical areas. They sleep during the day, often coiled up in the fork of a branch in the sun. At dusk they lazily unwind, undulating in the grass—and then suddenly seize their intended victim in their jaws. They are guided not by their eyes but by their tongue. They project it forward, picking up chemical particles from the environment; thus they locate and track their prey, explore the ground and find their way in the dark.

The female is oviparous. A single laying can include thirty to forty eggs, which the female broods—something quite exceptional in reptiles—with her body rolled into a nest around them and her head lying on top of them.

The python is one of the largest snakes

The green tree python (*Chondropython viridis*) blends into the leaves of the trees where it usually lies. Its tail, adapted specially for this habitat, is prehensile. Natives of New Guinea consider the python a tasty treat.

◀ Coiled in the shape of a nest around thirty to forty eggs, the female broods for about eighty days. This behavior is exceptional for snakes, which usually let the heat of the sun hatch their eggs.

The royal python (*Python regius*) of western Africa does not behave like a king. When threatened, it rolls up in a ball and hides its head in its coils. Though natives fear it, it is harmless. ▶

in the world, and the reticulated python is the largest of the pythons: it can be as long as thirty-three feet. It lives in southeast Asia, in the Philippines, and in eastern India, in dense virgin forests. It swims as naturally as it crawls. It sometimes strangles even pigs and dogs, but ordinarily it flees from man. At hatching time young reticulated pythons are already almost thirty inches long and weigh about three and a half to four and a half ounces. They grow about two feet a year until the age of five, and then their growth rate slows down to only one foot a year.

The Indian python, or *Python molurus,* is a water snake twelve to twenty feet long. In Ceylon, it hunts hens, geese, large frogs, and other reptiles. The rock python, the same size, lives in Africa south of the Sahara, in areas of dense vegetation. It hunts rodents, small pigs, antelope, and large birds. Finally, the African royal python is the smallest python of all, fifty to sixty inches long, and the most cowardly, too. When it is frightened, it coils up and hides its head in the middle of the ball which it forms. One rarely sees it adopt this comical position in the zoo, since by then it is too tame to really become frightened.

Macaws

The macaw, bird of sun and fire, dressed in colors of ruby and amethyst, turquoise and emerald, lives in the tropics of the New World, usually in the virgin forest but sometimes on the plains and savannahs. Macaws are long tailed parrots with naked, unfeathered patches around the eyes. There are dozens of species, some three feet long, others pigeon-sized. Like all parrots, including parakeets, their amazing plumage is made shiny by a fine, oily powder; and they have a croaking voice capable of startling modulation, which can repeat human words, and although the bird certainly doesn't understand them, can seem to deride them. Macaws use their hooked beaks to pick, clean, scratch, crush fruit, and crack nuts. With their feet, they catch food and bring it to their beak: roots, leaves, buds, berries, insects, and larvae.

Almost all macaws are monogamous, living as couples and in groups. They eat, drink, sleep, mate, and sit on their eggs together, though without much fondness for one another—beak fights are frequent. They lay their eggs in termite nests, tree hollows, or holes in rocks. A baby macaw, born without feathers, blind and ugly, remains where it is hatched for three to six weeks. Its parents feed it with food stored in their gullets.

Are macaws the most intelligent birds? Ornithologists are still debating this question, though they agree that macaws do have a startling memory and an enviable longevity: some live to be a hundred years old.

▲ The red and yellow macaw (*Ara macao*) is clad in clear red ruby. It lives communally in the impenetrable forests of Central and South America.

◀ A green-winged macaw couple (*Ara chloroptera*), a colorful and tender duo. Male and female macaws are faithful to each other for many years.

Breaking a nut, even a palm nut, is easy for the scarlet macaw. Its hooked beak is as strong as pliers. ▼

◀ The blue and yellow macaw is at left; beside it, the red and green macaw (*Ara militaris*) with its red helmet and turquoise-hemmed wings.

The glossy look is easy with the proper cosmetics: ▶ the macaw's feathers are covered with a film of shiny oily powder. This one is *Ara ararauna.*

The macaw's feet are used for climbing, sitting, and grabbing. The toes are paired, two in front and two in back. Here are two aras, the largest of the macaws (*Ara ararauna* measures three feet, ▼ *Ara militaris* twenty-six inches).

Weaverbirds

Among birds as among men, there are small artisans and great builders. Weaverbirds, which are city planners and perfectionists, belong to the second category.

The male is both mason and architect. He interlaces, weaves, knots, pads, and lines the nest with amazing skill. He anchors it to the branches, often above water, with bridges made of fibers and vines knotted together to prevent any chance of falling.

Each species adds its own variations. The African weaverbird builds a wall in its pear-shaped nest so that neither the eggs nor later the baby birds will fall out during a storm. The forest weaverbird (*Malimbus*) builds a long access corridor running the length of the nest from the opening, which is always located on the side or underneath. The sparrow weaverbird, which lives in continental Africa, builds a giant barbed nest, the framework of which is made of thorny mimosa branches and can be as long as six to nine and a half feet, and four and a half feet high. The fist-sized entrance narrows and lengthens into a tunnel. Inside this giant construction, sheltered from all enemies, there is often a veritable village. The sociable South African variety have chosen a covered town. Eight hundred to a thousand nests tightly squeezed together hang from the mimosa-clustered branches, the whole protected under the high arch of a common roof. Each year the weaverbirds add new nests, hung under the old ones. The protective roofing thickens continuously, and the whole construction becomes so heavy that it bends the solid trunks of the mimosa trees. "Streets" allow traffic among the nests, and there is an atmosphere of activity similar to that of a large human town during rush hour. All damages are immediately repaired by the males.

The female lays and broods three or four eggs twice a year, in the safe shelter

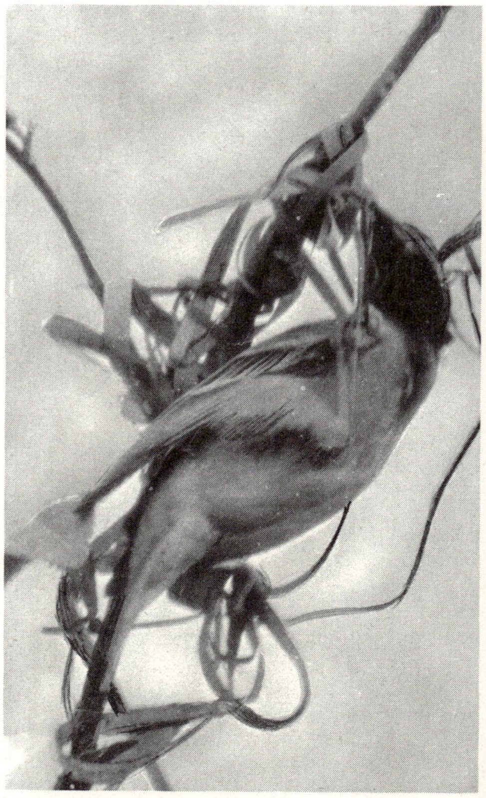

▲
Weaverbirds (*Ploceiodae*) are master builders that build their nests with incredible skill. They lace, weave, and knot highly varied materials, anchor their masterpieces in the branches with knotted cables, and connect them by hanging bridges.

The shady African acacias often serve to support the aerial villages of the weaverbirds. The nests hang from the trees like exotic fruit, without disturbing these antelope (*Oryx beisa*). ▶

The pear-shaped entrance is always on the bottom of the nest.

There is a funnel-shaped corridor at the entrance to the nests of some species, notably the sparrow weaverbird. A shield of grass and thick woven fibers encloses the nest of this weaverbird (*Ploceus vitellina*).

▲ Thorns often line the outside wall of the nest and make it even more strongly fortified. Balls of mud cement the construction.

of her house, village, or town. The baby birds, which she lovingly cares for, are five to seven inches long and covered with dazzling feathers, particularly the males.

Researchers have counted 263 different kinds of weaverbirds, most of them in the tropical regions of the Eastern Hemisphere from Africa to Australia, some in Europe and Asia, and one type in America. All feed on insects, larvae, grains, berries, and fruit. Their cone-shaped beaks shell the cereals; this characteristic is responsible for the red-billed quelea's (*Quelea*) well-earned title of predator. This species falls on grain fields by tens of thousands, devouring them. Poison or fire drives them away, but only to the next field. South of the Sahara, their large ball-shaped nests hang from the trees by the thousands. Millions of weaverbirds are born in them, and in their turn will gather in new swarms.

Branches bend under the weight of the constructions of the social weaver (*Philetairus socius*) that constructs giant communicating nests with mimosa branches and grass. The hard-working males maintain and constantly perfect their hanging covered town.

The Rhesus Monkey

In the wild, rhesus monkeys (*Macaca mulatta*) feed on fruit, seeds, and insects, and somtimes even aunch daring raids on plantations.

In India, the rhesus monkey is considered a sacred animal; in Europe, he is a circus entertainer with a barrel-organ; in America, he flies into outer space. But perhaps he is best known for the blood factor that bears his name: the Rh factor.

In 1940 K. Landsteiner and A. S. Wiener discovered a special agglutinogen in the blood of this monkey, which was named the Rhesus, or Rh, factor. This factor was found in the blood of 85 percent of the human population as well. Those who have this blood factor are referred to as Rh positive, and the other 15 percent as Rh negative. The blood type and Rh factor of a patient are now determined before any blood transfusion is performed, for a transfusion of Rh+ blood to an Rh− person can be fatal, since the Rh− individual would build up antibodies to the Rh factor. The incompatibility of blood factors between a parent with Rh+ blood and a parent with Rh− blood, starting with the second child, can cause mishaps in pregnancy or necessitate a total blood transfusion in the infant.

In the last ten years the rhesus monkey has become a space traveler. Strapped to his seat and covered with electrodes and measuring devices of all kinds, he has been sent into space to determine survival conditions in interplanetary rockets before sending men. We do not know how many monkeys have died this way. The best known was called Bonny. NASA sent him up in a satellite to study the effects of weightlessness on an animal as

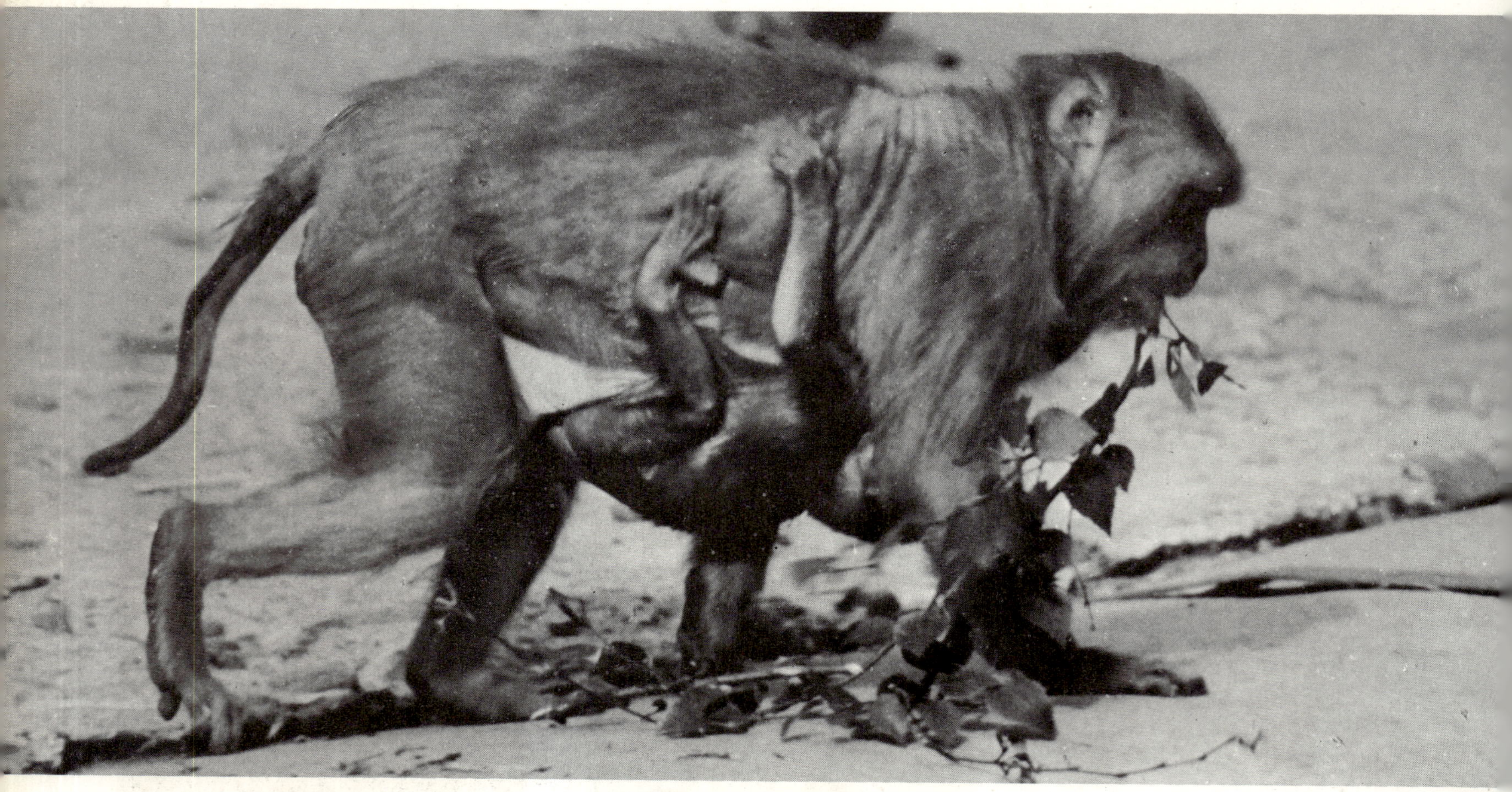

The newborn clings solidly to its mother's fur. The mother, always running or jumping about on all fours, carries the young everywhere with her this way. Doctors often use rhesus monkeys for laboratory experiments, and rhesus monkeys were sent up in the first space capsules to test life conditions in a weightless atmosphere.

Living in bands in forests, these intelligent animals are quarrelsome and fight loudly, argue, and bite one another. But the females are tender and attentive to their young.

similar as possible to man. Bonny stopped drinking and after two days lost interest in the games which were meant to test his attention by rewarding him with food rations. He died upon return to earth.

The rhesus monkey is approximately twenty to twenty-six inches long and eighteen inches tall. It belongs to the macaque group within the family Cercopithecidae, or Old World monkeys. Once scattered over a large part of Europe, the rhesus monkey is found now only in India, in the forests along the Ganges, and in the Himalayas at altitudes of up to 6,200 feet. Snow and cold do not bother them. In freedom, it lives in noisy bands of pillagers, leaping from tree to tree, running along the ground, jumping on rocks. It feeds on fruit, grains, insects, and sometimes even plans raids on plantations.

In captivity, its turbulent thievery and facetious good temper make it irresistible to its master—but only when it is young. Once it matures it becomes grumpy, irascible, and highly destructive. Its anger or irritation is manifested in changes of color: when the monkey is angry, its furless hands, face, and ears redden. At these times, it is wise to confine the animal or remove anything breakable from its reach.

Termites

"There is hardly any other creature so poorly armed by nature for life's struggle," wrote Maurice Maeterlinck. "And yet, with the help of that invisible element which in animals is called 'instinct' and in man called 'intelligence,' the termite has created its organization, made its dwellings impregnable, and provided for the future. It has multiplied infinitely, and has thus gradually become one of the most tenacious inhabitants and conquerors of our planet."

There are about fifteen thousand kinds of termites in tropical regions, and several others elsewhere. They live assembled in giant city-states of several million individuals, sometimes underground, sometimes in high round, straight, sugarloaf-, or mushroom-shaped mounds. And they always display that collective organization that conjures up terrifying science fiction visions of the future in which the individual's rank and form are determined by his social function.

The history of each new termite nest begins with a king and queen. When a termite colony reaches its demographic saturation point, a swarm leaves it. The individual termites that leave are ten times larger than those that remain, and different in shape and color. They depart at dusk while the birds, their predators, are asleep, and move in the direction of rain so that on landing they find workable soil. It is not so much a nuptial flight as an engagement trip during which partners choose each other. The newly constituted couples land one by one, and break off their wings. They look for a hole, in which they dig back to back with thin claws in the humid soil. They mate as soon as the first chamber is ready. The queen soon begins to lay, producing about six eggs a day, each less than a millimeter in diameter. The king and queen attend to the young together, feeding the larvae with secretions from their bodies. Licked untiringly, these larvae gradually grow, going through several successive moltings. After two months, they go to work alongside their parents. New generations of worker termites and soldier termites, both blind, are thus formed. The workers continue the construction of the termite nest, building its long rooms, ventilation shafts, store rooms, drainage compartments, and pools of underground water; they manage the complex and lavish care on the eggs and on the royal couple. The soldiers, who in some species have a nose-shapped append-

A giant mushroom in the tawny landscape of West Africa, this termite nest, several yards high, is the work of hundreds of thousands of worker termites, none of which is longer than half an inch.

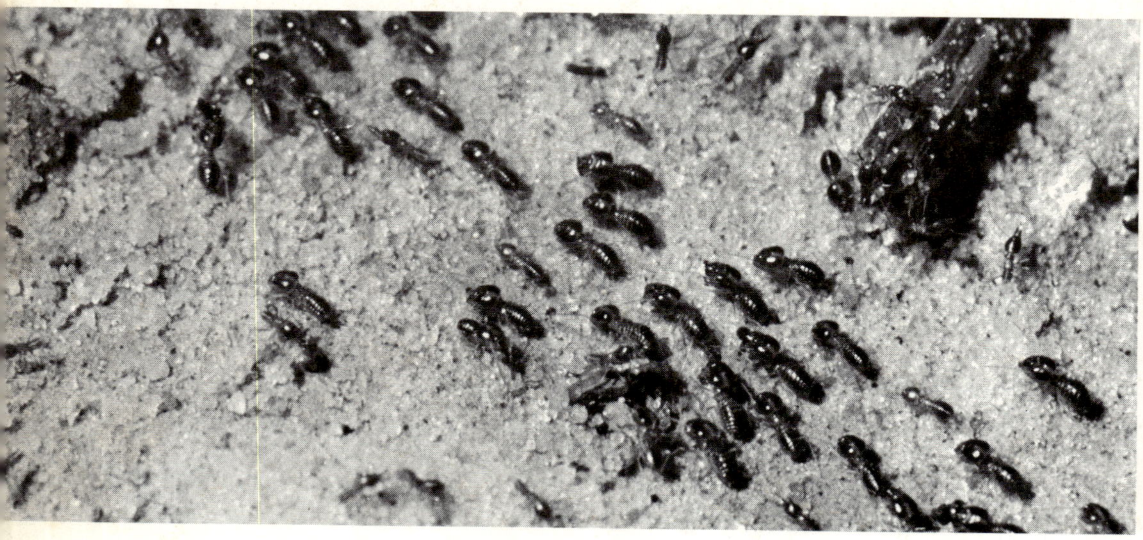

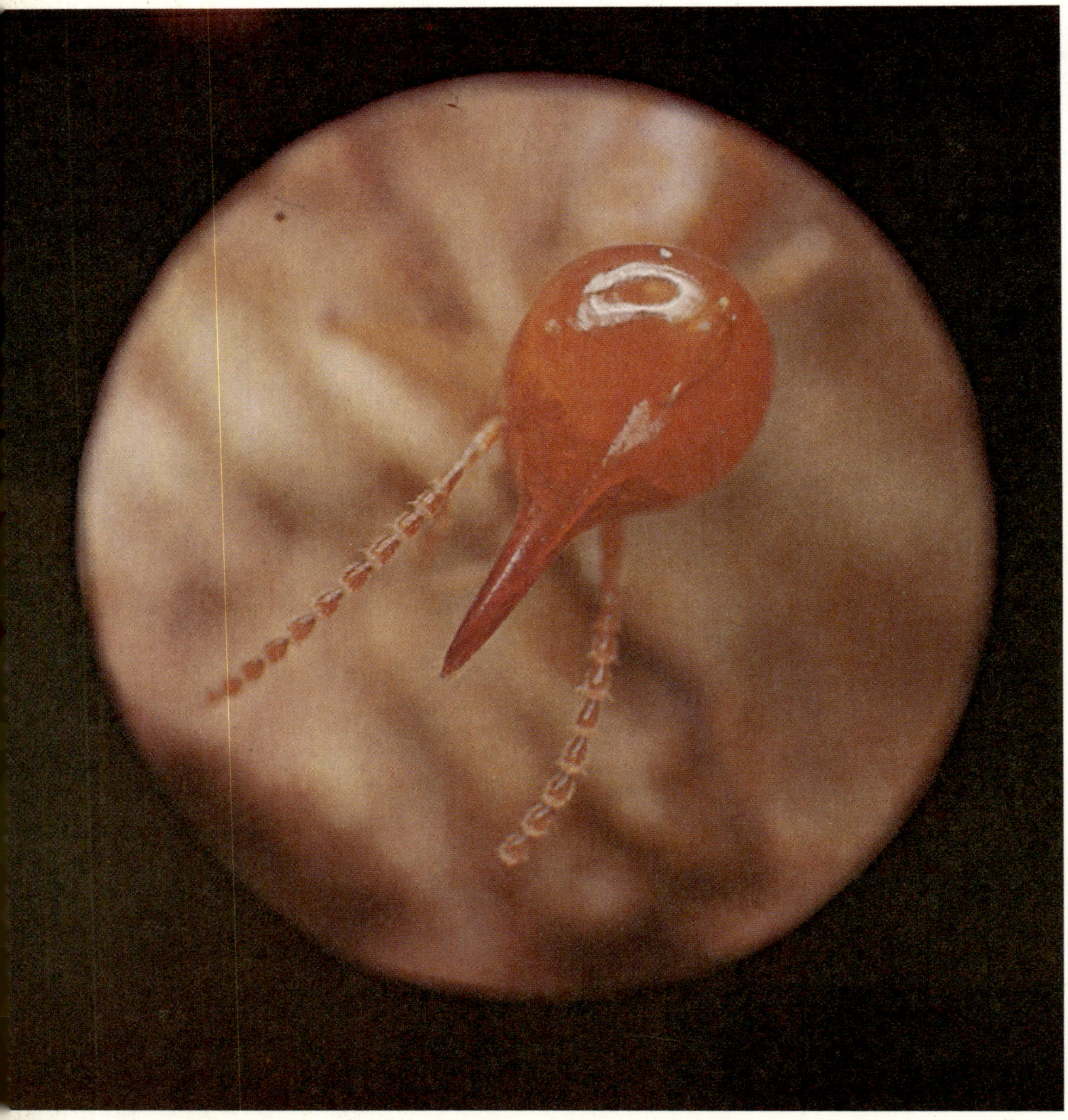

◀ Instead of atrophied mandibles, these soldiers (*Nasuti*) have on the front of their head a noselike appendage from which they shoot a defensive liquid.

Under the influence of a hormone that differentiates them, termites are specialized as workers and soldiers. These soldiers of the species *Bellicositermes bellicosus* with their enormous heads, defend pyramids twelve feet in height. ▶

◀ Under cover of night, worker termites transport their construction materials to the nest. During the day, they perfect and enlarge their underground city.

◀ Giant constructions are often supported by a tree. It is usually a dead tree, having been entirely consumed by its tenants, to whom it serves first as food, and then, in the form of digested waste, as construction material.

age that emits a sticky liquid, assure the defense of the nest.

Although among ants only sterile females undertake social tasks, both male and female termites capable of reproduction take part in work. Their roles are apparently determined by substances similar to hormones, which are continuously transmitted by reciprocal licking, allowing the balance among the categories of workers to be maintained. The royal couple, meanwhile, has only one duty: reproduction. They live secluded in the remotest room, from which they never again emerge, occasionally mating. The queen becomes a kind of egg factory, now laying them at a rate of eight to ten thousand a day. In her average life span of ten years she will have given birth to millions of offspring. She lies motionless, her little mate at her side—he is less than an inch long, while she measures three to ten inches in length and two inches wide. With her enormous abdomen she seems so formless that a close look is needed to distinguish her bright brown head, her eyes, and three pairs of legs at her front end. Surrounding her in a double ring stands the "royal guard": soldiers literally form a chain, head to head, jaws facing outward, ready to repel any troublemaker by main force. In front of them stretches the line of servants, according to an immutable ritual. They enter through one opening in the ring of guards to lick and feed the queen and her king, to collect the eggs and deposit them in incubators, and then leave through a second opening, always in or-

◀ The queen, the vital center of the community, lies in the hollow of this torn open nest. She is enormous: over three inches long and one to two inches wide, she is the mother of her millions of fellow citizens. Swarming around her are those who have just fed her and those who transport the eggs she lays to the incubators.

derly single file. Occasionally a sparkling bluish drop falls from the hind end of the royal abdomen, which is quickly inhaled by those workers nearest to it. The excitement with which this event is received leads the observer to surmise that the droplets cause a kind of drunkenness.

The normal diet of termites consists of decayed matter, paper, stalks, cereals, and especially wood—with a marked predilection for wood overgrown with mushrooms. Some termites raise these mushrooms themselves, in actual mushroom beds, fertilized with rotten wood and their own wastes. A termite colony in a house, silently consuming wooden beams and steps, gnawing away at balconies, can have disastrous consequences.

Usually made of chewed wood or earth digested in the intestine and mixed with saliva, the termite nests are as solid as concrete. In Africa and Australia, some reach a height of twenty-five, even forty and sixty-five feet. To maintain the same ratio, a man-made building would have to be almost a mile high!

Within these fortresses the termites are safe, sheltered from the birds, toads, and other insects that attack unarmed worker termites. Dynamite is sometimes necessary to destroy termite nests in Africa. The South American mounds, which are weaker and have a cardboard-like consistency, have a special enemy: the anteater. This predator tears the opening apart and makes a sumptuous meal of the teeming mass of termites.

But even in case of attack or accident—unless, of course, its millions of inhabitants are all annihilated—the termite colony fills in its gaps, and goes on. A swarm will leave it, to give rise to new colonies. Ad infinitum.

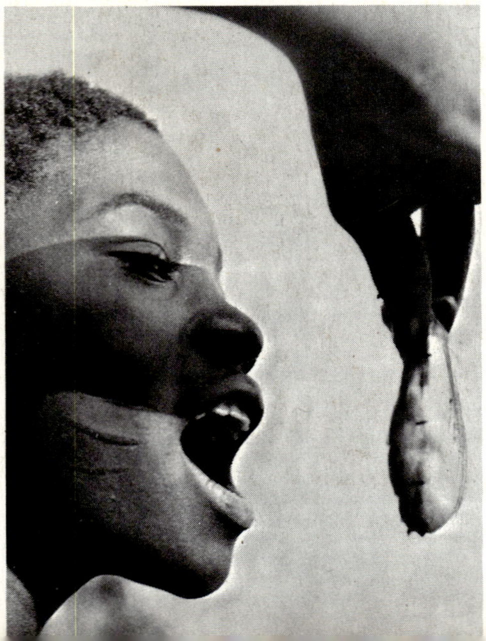

◀ A roasted queen termite is a native delicacy. According to its partisans, the worker too is delicious, especially when freshly grilled.

The little king, less than an inch long, never leaves the queen. He has only one function: fertilization. Through the years his mate has become totally disproportioned, with a gigantic abdomen distended by the stock of eggs, and a barely discernible head and back that have not ▼ developed.

Hummingbirds

The smallest bird in the world flutters and buzzes in the humid forests and river valleys of Central and South America, and even in high altitudes of the Andes. A few species are found along the East and West Coasts of the United States. This is the hummingbird, which drinks honey and nectar, hunts insects and spiders. It can be as small as two and a half inches, and it lays its tiny eggs in a nut-sized nest. But its intensity of activity is inversely proportional to its size: in one minute, it gathers honey from forty or fifty flowers, between stops executing a whole aerial ballet to the rhythm of two hundred wingbeats per second. It dips its thin beak into each flower, slips its long tubular tongue into the heart of the corolla, and sucks in the nectar.

During the day, the hummingbird is a constantly running motor; its heart weighs several grams, one-fourth of its total weight, and its lungs are more active than those of any other bird. Such constant energy outlays necessitate constant food compensation: the hummingbird would die of hunger at night if nature didn't allow it to "declutch." Each night the little bird becomes rigid, its body temperature drops, and oxygen exchanges slow down: it actually goes into hibernation. In the day it becomes active and dynamic again, and soars away.

▲ There are 319 species of hummingbirds (*Trochili*); each has a straight, curved, or even half-moon-shaped beak that allows it to plunge into the heart of its favorite flower. The length of the beak, as well as the shape, is adapted to need. The record is five inches, which is much longer than the bird itself. Here is the emerald-throated hummingbird (*Amazillia spec.*).

◀ As the hummingbirds' legs are short and weak, they are good only for perching. They cannot be used for running, but this is just as well, since in the hummingbird's tropical and subtropical habitat countless enemies would quickly devour it if it were on the ground.

Beating its wings two hundred times per second, ▶ the tiny bird buzzes continuously around the flowers it feeds on, without landing on them. At night, in order to rest, the hummingbird becomes rigid and enters a hibernation-like state.

4. Evergreen Forests

The Stag

As autumn comes, the silence of the forests of the Old World is broken at nightfall by a roar. Its head raised high under majestic antlers, the great red stag bellows with all its might: "I am master of all the doe." It is mating season, and the stag is defiantly summoning all rivals. If another stag takes up the challenge, a dangerous, sometimes fatal, combat ensues. The winner mates with the does, and the bevy is then reconstituted: the other males follow at a distance. In May or June the fawns are born: one or two per doe, graceful, fragile, and helpless. The white spots scattered on their brownish-red coats at birth gradually disappear. At the age of six months, two bony two-inch-long protuberances appear. Between ages one and two, these grow into eight-inch spikes, and the male is now called a brocket. From then on, two tines, or points, will grow on the antlers each year. In hunting terms, it becomes second, third, and fourth head. A thick layer of skin shot through with blood vessels covers the antlers. In June, the stag sheds this velvet by rubbing its tines on tree branches. It goes thus "bare-headed" until February. Then, as though the antlers were dead branches, they fall off and new ones begin to grow. A stag's age cannot be determined exactly by the number

The great red stag (*Cervus elephas*) permits no sharing of his harem. But it socializes with the does only during the mating season, normally preferring male company.

Winter is a hard time for red deer. Hungry, they become fierce, and only in the most dire need resign themselves to attempting the long trek to the food that forest rangers set out for them. Even then, the snow is often too deep for them to reach it. ▶

of tines, for physical development, including antler growth, varies according to the quality and supply of food. Generally, ten- to fourteen-year-old stags have the most powerful antlers. They can be as long as four feet to four and a half feet, measured along the curvature, and weigh about thirty-three pounds. Hunters and riders have found antlers of massive old stags with twenty-four tines or more. One semicaptive stag was found to have sixty-six tines, weighing over forty pounds.

When a stag is described as "serious, imperial, frowning," it is always the red stag, or *Cervus elephas*. Inhabiting the European forests, it has been the source of countless tales, in which it is often a king or prince under some sort of spell. The most powerful red deer, those of eastern Prussia, can be as long as seven and a half feet and stand as high as five feet. The smallest, under two and a half feet, live in Corsica. However, the Barbary red deer, like its cousin in the Atlas mountains, is almost extinct. Related species are also found in central Asia, Asia Minor, and in North America, where they are called wapiti, or elk. Imported red deer have learned to acclimate and reproduce in the peat bogs of Scotland and in Australia.

Red deer live an average of seventeen years. In Europe, red deer are protected by hunting regulations. They live on grass, plants, leaves, and bark. Whatever damage they may cause to their environment is accidental, occurring in winter when they are racked with hunger and prevented by thick snowdrifts from reaching their feeding areas. Except in pre-mating duels, the stags' antlers serve no purpose and seem rather to get in their way, preventing them from getting into the undergrowth to feed. But forest rangers claim that stags are precious helpers to them, as they rid the trees of dead or useless branches.

▼ In the beginning of summer, the stag sheds the velvet that covers the antlers.

▲ In autumn, the king of the woods approaches the does. Then his cry rips the silence of the night, as if all creative forces, and all instincts of a free life, were exploding through his voice.

◀ In the spring, the fawns stand on their hind legs to reach the tender fresh leaves.

Master against bachelor, this is the duel for possession of the female. Combat is always furious. The adversaries wound and tear each other, and sometimes the contest ends in a death. ▶

Old World Goshawks and Sparrow Hawks

▲
The cautious northern goshawk (*Accipiter gentilis*) avoids man, who hunts it because of its habit of raiding chicken coops. Its eyrie is always hidden in the deepest and highest coniferous forests, far from villages.

◀ Its small cousin, the European sparrow hawk (*Accipiter nisus*), is bolder, and is found everywhere.

The griffin of mythology—with its leonine body, horselike ears and eagle's wings and head—was the most terrifying of all the animals. Goshawks and sparrow hawks are hardly more reassuring, with their icy pupils, cruel beaks, and pointed talons. Though they are born killers, they spare their victims the agony of a prolonged death. When they swoop down on their prey, claws open, death is instantaneous. They use their beaks not to attack, but to pluck or skin the victim before devouring its flesh.

The northern goshawk, which inhabits Europe, Morocco, Asia, and North America, is bronze-colored; it is fast, agile, and determined. The female is slightly larger and heavier than the male, twenty-eight inches and two and a half pounds to his twenty inches and one and a half pounds. It hunts squirrels, pigeons, hares, rabbits, and other birds of prey such as its smaller relative the sparrow hawk, the buzzard, or even the sparrow, pouncing on them suddenly after a long patient watch.

In antiquity, the goshawk, like the falcon, was trained for hunting. But its habit of stealing poultry earned man's animosity, and today the goshawk is cautious around man. Its eyrie is difficult to approach, since it is a veritable fortress anchored on a tree, between a branch and the trunk or hidden in the solid fork of a coni-

fer. It is even harder to observe the male's courtship of the female in the spring. He flies around her in perfect circles, coming down in nosedives like a fighter pilot. At the end of April or the beginning of May, the female lays two to five eggs, and thirty-six days later the baby birds hatch. The parents feed them for five or six weeks, and then the young birds leave the nest. Goshawks become sedentary and do not migrate when they are older, but when young, they leave the rigors of winter behind to fly to the sunny south.

The European sparrow hawk inhabits all of Eurasia and northeast Africa. Physically it resembles the northern goshawk, though smaller. The female is only fifteen inches long and weighs half a pound, and the male twelve inches and a third of a pound. But the sparrow hawk differs from the goshawk behaviorally. It prefers to nest in conifers, near gaps in the branches that permit easy access to the eyrie. There it stays, surrounded by the remains of sparrows, finches, pigeons, jays, and crows, all its favorite prey, which it grabs in midflight.

Each clutch of sparrow hawk eggs consists of three to six young, which hatch after thirty-two days of gestation and leave the nest at four weeks. The male kills the prey birds and takes them to a spot where the female, who feeds the young, comes for her share. If the female dies, the babies die too, and the father finds a new mate.

Avid, the watchful bird of prey pounces on its victim, a squirrel. The panicked rodent tried to escape by rushing down the tree trunk, then dived into the air, legs spread far apart and tail acting as a rudder.

The European Red Squirrel

A bright red sunbeam bolts from branch to branch, up and down, here and there, bravely approaching a human stroller and then rapidly fleeing, as much afraid as mischievous. The forest squirrel is always playing hide and seek.

European red squirrels are found from the far north to the Mediterranean, from Ireland to Asia. From Rumania to the Alps, the most common variety is red with a white belly, a long bushy tail and, in winter, thick whiskers that cover its ears. In cold weather the squirrel dons its winter outfit: a longer, shinier, grayer and lighter-colored fur. It bustles around endlessly, searching for seeds, fruit, nuts, beechnuts, buds, bird's eggs, and even stolen baby birds, which it munches sitting up, bringing the food up to its mouth with its front paws. It is completely at home in the trees, head up while climbing up, head down when climbing down, its sharp-clawed toes planted solidly in the bark.

This diurnal animal lives socially in a family. It builds a round nest filled with moss, leaves, and hay, perched in the fork of a large tree. Two, or sometimes three, times a year the female gives birth to two to six baby squirrels, after a gestation period of thirty-six to forty days. The mother nurses them, cares for them, and when she must leave the nest for a moment fills the entrance of the nest with a wad of leaves or moss.

In winter, the squirrel stays mainly in the nest. It rolls up in a ball, covers its body with its tail, and sleeps. When the temperature becomes warmer, it wakes up, rises, runs to one of the caches which it had

◀ Climbing isn't easy when one is very young. But with mother's help, one succeeds.

▲The fastest way to get around is to leap, all four legs stretched out to break and cushion the fall.

"Red as a squirrel" is a common expression in some parts of the world. But in the mountain forests of central Europe, most species are black.

filled with nuts—beechnuts, acorns, or pine cones—in the fall, and has a snack before going back to sleep until it is once again awakened by hunger. This lazy behavior continues until the warm weather has really returned.

The European red squirrel is dangerous only to baby birds, but it has two relentless enemies: the otter and the goshawk. Flight is its only weapon when threatened by either one. When it is brought to bay, in order to escape it leaps into the air, head down, tail in the air, and legs spread far from its body to break the impact of the landing.

Titmice

The Bible says that God will feed the birds of the air, and the titmouse lives according to this promise, happy and completely unconcerned with tomorrow. It is all delicate grace and airy lightness.

Man protects and helps the titmouse as much for its beauty as its usefulness. Titmice are internationalists, inhabiting the forests, parks, and gardens of all continents. The Germans hang "titmouse wreaths" of food in their country windows. French bird-lovers leave it perches and food reserves, which it uses up in the winter. For the titmouse is more threatened by starvation than by the cold: if it is without food for twenty-four hours it dies. It needs incredible quantities of food—out of all proportion to its weight of a third of an ounce—for its perpetual activity as it flies from tree to tree and climbs from trunks to branches. It kills vast quantities of harmful plants and pest animals. No insect, larva, or chrysalis can escape its notice when it searches the cracks and barks of trees. One family of titmice alone (including seven babies) consumes three hundred pounds a year. Oily seeds, such as hemp or sunflower seeds, which it opens with its small short beak, complement its diet.

A sociable busybody and a little bit quarrelsome, the titmouse has a great sense of family life. It visits not only other titmice, but its relatives the wren, tree-creeper, pitta, and woodpecker as well; and especially in the winter, it flies with them.

Most titmice seek a natural nest at hatching time, a well-sheltered hole or a hollow abandoned by a woodpecker, or a man-made perch in a garden or forest. Some, though, build their nest themselves, and do it quite well. Thus the long-tailed tit, which lives in Europe, Siberia and Japan, builds an oval nest, entirely enclosed except for a tiny exit hole, in thick hedges or piny bushes. It lines the nest with moss, lichen, spider webs, and interwoven plant fibers, and camouflages it under small bits of bark. The penduline titmouse (*Remiz*), found all over eastern and southern Europe and as far as central Siberia and China, hangs its pear-shaped, corridored

◀ This Old World blue titmouse (*Parus coeruleus*), its beak wide open, can already fly but cannot yet find its own food, and so sits on a branch while the mother feeds it.

With its wings acting as a brake, the great tit (*Parus major*) is landing. This photograph was taken with an electronic flash. The wings, closely hugging the body, spread out like a fan. ▶

▲
The skillful marsh tit (*Parus palustris*) weaves its nest of green mosses, hare's fur, and down.

▲
The shy and pretty blue titmouse hides its small, well-upholstered home in the hollow of a tree. Two or three times a year the female lays about a dozen eggs, which thieving squirrels lie in wait for.

◀ The clever marsh tit has several homes: its nest, in which it lives and broods once a year, and secondary homes, which are nooks or cracks in trees that it slips into at night.

The hungry titmouse eats great quantities of food, about 75 percent of which are caterpillars. The great tit eats about 130 a day, each one and a quarter to almost two inches long.

home above water. The male does the work, using downy willow fibers or wild hops fibers from poplars. The female's sole concern is comfort, while the male's building instinct is so powerful that he builds several nests in a row, one right after the other.

The most common titmice in Europe are blue titmice, great tits, and marsh titmice, which are balls of gray feathers with yellow cheeks.

Titmice grow and multiply easily. Two or three hatchings a year each give rise to five or six young, sometimes twice as much. Jays, woodpeckers, crows, and thieving squirrels, which are all bird-egg and baby-bird-eaters, counterbalance this fecundity. Without them, the world might be overrun with titmice.

Ants

Ants have been admired for their unflagging determinations, while La Fontaine gave them their reputation for stinginess in his famous fable. Useful because they destroy harmful plants, they are red, black, or brown, giant or tiny, mean or peaceful. They are everywhere. The uniqueness of their mode of social organization, the artfulness they constantly display in improving their living conditions, never cease to intrigue entomologists, from Maeterlinck and Fabre on.

Ants live in collective states in which each citizen performs one well-defined function all its life, without the slightest chance of promotion or change. In the species in which the males and females have wings, the history of an ant's nest begins with a mating flight. Although the laws governing the process are still unknown, at a certain point each ant community produces sexually differentiated individuals, suited for reproduction. Upon adulthood, these fly away in swarms of thousands. Mating occurs in the air, and immediately afterward the male dies and the female lands. She is now a queen, and she must begin to build her kingdom. Her wings drop off, and she immediately sets out to find a spot in which to found her state: perhaps a hollow in a tree stump, a hole in the ground, or a rock to slip under.

Red ants (*Formica rufa*) protect their subterranean dwellings, built in humid areas and shaded by conifers, under a dome of leaves and dry needles up to five feet tall. The dome insures temperature control of the ant hill and is a sort of summer home where the worker ants place the chrysalises in the heat and light of the sun. *Right:* A battle between the defenders of the fortress and the invaders.

Well-barricaded in her home, the queen begins to lay. She herself cares for the first generation. The larvae that emerge from the eggs will become chrysalises, and then workers. They immediately begin to do the housework in the ant colony, and then they care for the subsequent generations. Meanwhile, the queen attends to the proliferation of her subjects, and in a few years will have produced a million.

Each ant colony has one or several queens, and many sterile female worker ants: the masses. This proletariat constantly maintains and repairs the nest, cleans and tidies up, brings in stores of food and prepares it, defends the state by biting and spraying any attacker with a corrosive liquid, and, most important, cares for the young. Worker-ants help them complete their metamorphosis by breaking their cocoon with their jaws so that they can emerge, then lick the larvae to keep them from drying up. They drag the chrysalises up to the heat and light of the sun in the day to help their hatching, and at night take them back down into the safety of the city depths, and will evacuate them if there is danger of flooding. Certain species, such as the red forest ants, add a summer house to their subterranean dwelling, a pine-needle and dried-leaf dome that can be up to five feet high. This roof insures the thermic isolation of the ant hill, and during the day this chamber is the nursery for the chrysalises.

An ant lives fifteen to twenty years. It is blind or nearly blind, but it is guided by smell, taste, and touch. It can carry over fifty times its own weight. This is the

▲ A world's record: the ant can drag an object fifty times its own weight. These red ants are transporting white pebbles with which they will build a terrace at the entrance to their nest.

Above, right:
Danger: a hungry slowworm, a kind of legless lizard, is gliding into the ant hill.

Opposite:
The red ant has the advantage: it has thrown its adversary to the ground and is biting it violently.

A red ant has captured a bumblebee ten times its size. ▶

equivalent of a man lifting three automobiles at once, 7,500 to 8,500 pounds.

Though it is impossible to describe in just a few sentences the singularities of the ant world, a few of the most spectacular are worth mentioning. Tropical tree ants use a certain number of their chrysalises as tools, employing them as weavers to sew together the leaves that form their homes in the trees. Driver ants and army ants have no fixed home and wander in huge armies, attacking everything alive, domesticated animals, fowl, or man; they

▲ Sharing food comes naturally to ants of the same species. Here a worker ant is offering a piece of nut to a new arrival.

◀ Animal husbandry: the red ant on the left is using its antennae to "milk" a green fly in order to gather its sugary secretions. The worker ants are enthusiastically participating in the operation.

cross bodies of waters by hanging on to one another until they form a live bridge over which the rest of the troop crosses. Mushroom ants cut great quantities of fresh leaves which they stuff in their ant hill for a mushroom bed; they feed on the mushrooms, which supply them with a protein they require; on its mating flight, the queen takes along in her gizzard a dowry of a cutting which will allow her to start the farming indispensable to the subsistence of her state. Slave-making ants steal chrysalises of other species which they use as their slaves. Pastoral ants build walls that reach the plants on which green flies feed, and thus have veritable herds which they milk, as farmers do cows.

Finally, the honey ants of America, Africa, and Australia have a bizarre custom: fifty to three hundred workers are "fattened up" with sugar and leaf and flower extracts, until their abdomens are as large as peas. They are then hung up, like live honey jars, in the food storage rooms. They are used to feed the other workers until, drop by drop, they are emptied and die.

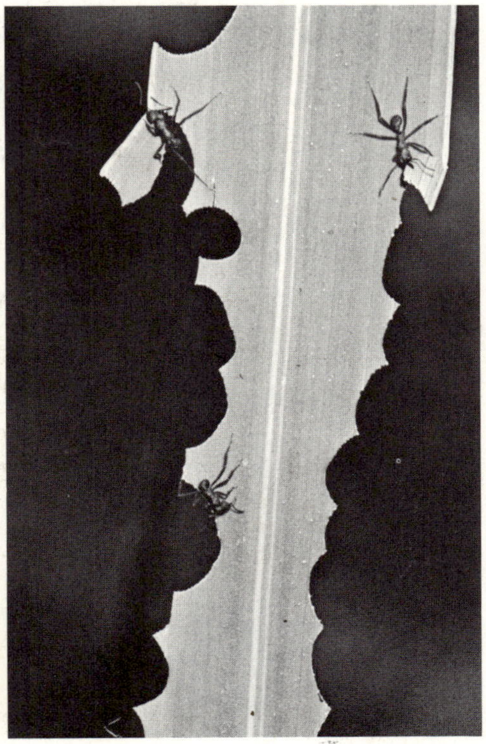

Agriculture: mushroom ants (*Attini*) are cutting fresh leaves into small pieces, which they will carry to their nest and make into a mushroom bed.

Maneuvers: driver ants (*Dorylini*) have no permanent home, so they roam and pillage. Here a giant convoy is advancing in the virgin forest of ◀ the Cameroons.

Spotted and Green Woodpeckers of the Old World

The young of the great spotted woodpecker (*Dendrocopus major*) are voracious eaters. Their parents feed them thousands of insects, which they untiringly gather in the bark of trees.

Some birds sing to express their love and to announce the borders of their territory. But the spotted woodpecker prefers drumming. In the spring, its beak hammers on dry twigs at the rate of ten beats per second. The female then comes running. Spotted woodpeckers, and their green woodpecker relatives, can always be found by the sounds of their loud activity. The name "woodpecker" is a literal description. This bird uses its long straight beak, which is as hard as iron, as a worker uses his drill, to bore into trees to install its nest and the hollows in which it sleeps; it can split even the hardest bark in order to pull out larvae and insects. Because of the woodpecker's fine sense of hearing, as it pecks it can find the tiniest crack in the wood, revealing the presence of an insect. The woodpecker extracts the insect with its tongue, a tool perfectly adapted to its purpose. The tongue is thin and long, one and a half times the length of the bird's head, and its surface is hornlike and covered with tiny hooks. The tongue investigates the crevice and spears the larva secreted there. At rest, the tongue is rolled up far back in the head.

In winter, when the insects are gone and there is a scarcity of food, woodpeckers also eat seeds and fruit. They skillfully catch the nuts, acorns, or pine cones in cracks in the bark and open them with their beaks. Sometimes they even risk raids on beehives.

Both the mother and father brood and feed the four or five baby birds that are born in the spring and that rapidly become independent.

The green woodpecker (*Picus viridis*), a carpenter bird, digs its nest in tree trunks by using its long iron-like beak as a pair of scissors. It devours enormous quantities of insects, and has a definite ◀ predilection for ants.

Crossbills and Black Woodpeckers of the Old World

The crossbill (*Loxia curvirostra*), which likes conifer seeds, has a scissors-like beak which it uses to remove the splinters from pine cones and to reach the oily seeds. It divides this food equally among its family.

While for most birds in Europe and North America winter is a time of scarcity, for crossbills it is a time of plenty. Indeed, winter is the time these birds find great quantities of their favorite food: seeds from coniferous trees. Crossbills are rather like vagabonds, without a fixed territory: they fly from region to region in groups, stopping wherever there is food. Having eaten, they are once again on the move.

Again unlike most birds, crossbills mate and brood in winter, too. The female settles comfortably on the eggs, deep in a moss and lichen nest well hidden in the trees, safe from snow and storms. In only fourteen to sixteen days the baby birds break out of their shells. When they hatch they are covered by a thick down that pro-

◀ The indecisive black woodpecker (*Dryocopus martius*), the largest of the European woodpeckers, pecks several successive holes before deciding where to live. These half-finished constructions are immediately taken over by other smaller bird species, which gladly move in.

tects them from the cold. At three weeks the bill crosses, and the young can feed themselves.

The black woodpecker, which is the largest European woodpecker (eighteen inches long), works for others. Before choosing the final location for its nest, it bores several holes in a row, which it then abandons. These become ready-made refuges for other, smaller birds. In the same way, when it discovers larvae in the trunk of a tree, the black woodpecker moves on before the supply is completely exhausted, leaving food for other birds.

▲
An expert carpenter, the red woodpecker digs a purse-shaped hole where it lays its eggs. Even the strongest wood cannot resist this bird, which is only eight inches long.

5. High Mountains

The Brown Bear

▲
The first walks and the first pranks. At the beginning of summer, the female brown bear (*Ursus arctos*) allows her cubs to accompany her on expeditions, but the waggish and curious little bear pushes its nose into every hole on the way. The mother's reprimand is a slight tap.

Helpful or harmful? Bold or shy? Good-natured or sly? There is no agreement about the brown bear, and everyone is partly right: it is one of the most disturbing of animals. Each bear has its own distinct and unpredictable personality, and its behavior can suddenly change, with no warning alteration in facial expression. Its special talent is probably improvisation.

The brown bear, which is heavy, thickset, squat, careful, and not very sociable, walks with its rocking gait across its territory, the forests of Europe, Asia, and North America. In Europe, the great brown bear is found in the Pyrénées, the Abruzzi, the Tyrol, the forests of Scandinavia, Poland, and Russia, and in several of the Balkan countries. The Alaska kodiak, the largest carnivorous animal of the world (it can be ten feet long, four feet wide, and weigh 1,100 pounds), is one of the brown bear subspecies, as is the somewhat smaller North American grizzly, which is usually under seven feet long and 570–700 pounds.

The bear usually feeds peacefully on plants, mosses, berries, and mushrooms, or on small animals such as beetles, and happily raids beehives for honey for dessert. But this pacific vegetarian sometimes has a craving for meat. It then becomes a killer and attacks foxes, does, and even stags with its great paws. In the salmon spawning season, it lives on fish: it stations itself at good spots and catches the largest salmon with surprising patience and skill. Livestock that are guarded by dogs are relatively safe from the bear, who doesn't actually fear the dogs but is driven away by their barking. Like most wild animals, the bear generally flees man, facing him only when directly threatened by him. At that

◀ Standing on its hind legs, the Alaska kodiak (*Ursus arctos middendorffi*) is a threatening sight. Weighing 1,100 pounds and standing ten feet tall, it is the world's largest carnivore. Even so, this great bulk is capable of astonishing agility and climbs up trees like a squirrel to eat honey stolen from wild bees.
▼

A mother first and foremost, the female shows no interest at all in the opposite sex during the year she cares for her cubs.
▼

◀ The Kamchatka brown bear of the Soviet Union and the Alaska kodiak are experienced fishermen, waiting for the salmon to swim upstream and then catching the largest. The two giant cousins like the fatty fresh fish.

The brown bear is a tireless swimmer, comfortable even in the most dangerous rapids, where it remains for hours, waiting to catch large fish skillfully with its paws.
▼

point, a good rifle and the ability to use it are necessary.

For a brief time in summer, the male loses its usual reserve and becomes more active in order to attract and seduce the female. However, his advances are sometimes rejected if the female is caring for cubs or is old. If she accepts the male, she gives birth to two or three blind, furless, rabbit-sized cubs. The birth takes place in the winter dormancy period, during which bears take no food, sleeping in their sealed-off holes. The mother bear prepares for the birth by carefully lining her cave with moss, branches, and leaves. She picks up the newborn cubs with her paws and snuggles them under her warm breast. Their fur soon grows, and at about one month their eyes open. But they don't leave their mother's warmth for ten weeks. At three or four months, the cubs begin to gambol about, under their mother's watchful eye, and at six months they begin to accompany her on excursions. At autumn's end they prepare the shelter in which the young will doze alone all winter, near by, and they prepare to leave her.

Bears are inscrutable and disconcerting, and in the wild are nothing like the inoffensive "Teddy." Trainers claim that even the young have unpredictable reactions and are at times dangerous. ▶

The Alpine Marmot

For a long time, the Alpine marmot was a part of urban folklore. They were trained by Savoyard chimneysweeps to climb down chimneys and thus detect chimney fires; and at the turn of the century they brought city dwellers—nostalgic for the countryside they had left behind—a whiff of mountain air.

There are no longer any marmots in the towns of Europe. But in their original regions they still lead their peaceful group life, in high altitudes in summer and in the valleys in winter. Marmots are found in central Europe, northern Asia, and North America. The Alpine marmot lives, as its name indicates, in the Alps, as well as in the Carpathians, the ranges of Russian Turkestan, and as far east as the Altai and Tien Shan mountains and the Kamchatka peninsula. But there are no marmots in the Pyrénées.

The life of this rodent, in Europe the second largest after the beaver, is unalterably linked to the rhythm of the seasons. In the spring, after a long winter's sleep, the mating season comes. The normally timid and peaceful males suddenly become aggressive. They become veritable "bears"—as hunters call them—chasing the "cats" (the females) and angrily provoking all rivals. Then calm returns, and the marmot family, made up of two or three members, or the entire tribe, a combination of several families to-

▲ Summer is the favorite season of the Alpine marmots (*Marmota marmota*). Moving in colonies, they climb to the snow line, where they build their summer homes. Here they do nothing but eat well in order to store up reserves of fat for the winter.

◀ Friendship has its rituals: marmots rub noses in greeting.

Caution is always wise, even when on vacation. The ever-watchful marmot sits up watching its surroundings. At the slightest suspicious sound, it lets out a piercing cry and the whole tribe retreats to the shelter of the burrow. ▶

taling thirty to fifty animals, migrates to their second homes. These summer burrows, built at the snow line at altitudes of 6,500 to 10,000 feet, have long fortified corridors and several emergency exits. And, just before man's haying season, marmots too make hay. They use their teeth to cut the grass, blade by blade; they turn it over and over like skilled haymakers so that it dries completely; then they gather it together in piles and carry it in their mouths, bundle by bundle, to their hole. The hay will serve as padding, especially to make the nursery comfortable: this room, where the mother gives birth, is in a side corridor of the burrow. The young are born in litters of one to seven, in the beginning of June, after a gestation period of thirty-three to thirty-six days. At three weeks they can see, and by the end of June they can fend for themselves, and they leave the burrow and join the adults.

Then in summer the easy life begins. The marmots do nothing but eat well, thus accumulating reserves for the winter. But they always take certain basic safety precautions. Marmots never go far from their burrow, and are constantly standing up on

The marmot's long trek from the valleys to the ▶ mountains is fraught with danger. It advances cautiously, all senses alert.

their hind legs, like a dog begging, in order to keep an eye on their surroundings. When a marmot sees danger in the distance, it lets out a piercing cry to warn the others, and they all run for shelter. In areas where marmots are numerous, one can see a vast collective burrow with special lookouts, along with the smaller individual homes.

The "summer vacation" is over at the end of August, and marmots must think of returning home again. The round ball-like marmots leave the heights, the world of dwarf firs and mountain pastures, to return to the valleys that in winter will be free of heavy frost. In September and October marmots live in prepared refuges, whose entrances they barricade with dirt, stones, and hay. Then they snuggle up to

▲ The prone position is always safest for the marmot. When it is disturbed by a noise, or by the cry of an eagle or a jackdaw, the marmot flattens itself along the ground.

The entrance to the bobak's (*Marmota bobak*) burrow is marked by a pile of sand and rocks. This close relative of the Alpine marmot lives in the steppes of eastern Europe. It always leaves a few tufts of grass to quickly hide beneath in case of immediate danger.

one another and settle down for their long winter's sleep of six to seven months. They breathe only every three or four minutes, their pulse rate drops from two hundred beats a minute to five or six beats a minute, and their body temperature lowers to 37° F. By spring, they have used up the fat stored up in the short good times. They have lost half of their average weight of nine to sixteen pounds, and they are no more than flesh and bones. But the weather is beautiful, they poke their muzzles out of their burrows, and the cycle of life begins again.

The Golden Eagle

A giant shadow is often seen over arid heights, lonely forests, rocky cliffs, isolated plains. This is the shadow of the golden eagle. It is the strongest of all birds of prey—with a wingspan sometimes exceeding seven feet—and one of the noblest and most dangerous as well. Quail and pheasants, marmots and hares, foxes, roe deer, and fawns must all beware of its talons.

In Europe, the golden eagle is found in the Pyrénées, the Apennines, the Balkans, the Alps, and in Scandinavia. It is also found in Siberia, in eastern Asia, in North America and in North Africa. It lives far from man, in high tranquil and secluded valleys. Once it has chosen its home, it never leaves it. The golden eagle does not migrate with the seasons.

The male is up to twenty-eight inches long, and the female thirty-four inches. These birds live in couples for years—the average life span is thirty to forty years—hanging their enormous eyrie from a large tree, a rocky spur, or in the hollow of a rock. The female lays two eggs, which hatch after forty-four days. Baby eagles can see at fifteen days and fly at three months. But the law of natural selection is operative: often only the stronger of the two completes development, because as it struggles to receive all the food brought in by the parents, it pushes the weaker down into the bottom of the nest where it dies. When the survivor is strong enough, it leaves the eyrie to try its wings. It reaches sexual maturity only at age five or six, and then flies off to establish its own territory.

In Europe the golden eagle was long trained to hunt. In central Asia it is still used to hunt wolves and the Saiga antelope. But all the stories of fights between men and golden eagles, or children stolen by golden eagles, are only fairy tales. The golden eagle fears man as its worst enemy, and always flees him.

The eyrie of the golden eagle (*Aquila chrysaetos*) is an impressive construction. Sometimes as tall as five feet, and always inaccessible, it is a rigid platform on a sheer cliff or at the top of a tall tree. The male brings the food. The female empties his gullet and distributes its contents to the young.

The Great Horned Owl

In the dark forest, two round, intent eyes pierce the night like headlights. The great horned owl, the largest and rarest of the owls, is watching. Once found throughout Eurasia and North Africa, today it has practically disappeared from some parts of Europe, such as Germany. To protect these regal birds from extinction, some are now raised in captivity and then liberated at adolescence to live in freedom with their relatives.

One must penetrate lonely, difficult-to-reach mountain forests to see this nocturnal bird. It perches in rocks, hollowed-out trees, eyries abandoned by other birds of prey, and sometimes even on the ground. It only rarely hoots. It hides during the day, hunting only at dawn and dusk. Its favorite prey are rabbits, hares, squirrels, and small mammals, as well as some birds such as the crow and the grouse. It uses its beak to dismember its victims and pluck their feathers or fur. When it is full, it carefully puts aside the leftovers for its next meal.

At the end of March or beginning of April the female lays two or three eggs, each after a long interval. She broods them for thirty-one days, then tends them seven to ten weeks until the baby owls can fly. The mother and father bravely defend the babies against enemies, though they are powerless against fratricide: the firstborn often devour the youngest.

▲ The shiny, immobile eyes of the great horned owl (*Bubo bubo*) pierce the dark of the night. The eyes alone make up one third of the total weight of the head.

The young are a fluffy white; often the older ▶ siblings will eat the youngest.

Chamois and Ibex

▲ The shy Alpine ibex (*Capra ibex*) doesn't stray from the herd, whose territory lies between the timber line and the glacier zone.

If you were silently to inch up to a mountain crag where there were only rocks and bushes, you might have a slight chance of glimpsing a picture postcard scene: an immobile chamois, head upright, balancing on a rocky spur. But the tableau comes to life if the animal's meditation is disturbed by the slightest sound. The chamois runs off so fast that its fleet hoofs are already in the air when the rocks they dislodged begin to roll down to the valley.

The confident chamois (*Rupicapra rupicapra*) leave their high territory at dawn to graze in the ▼ valleys.

Though chamois are not very long (about forty inches) nor very tall (twenty-seven to thirty inches), they are wiry and graceful. The long hairs of their bearded chins, about eight inches long, are of the same texture as the hair on their back. Though the species is becoming extinct, they are still occasionally found at high altitudes in the Alps, the Pyrénées, the Caucasus, and the mountain ranges of central Asia and Asia Minor. The cloven front hooves of the chamois can be spread widely apart, thus allowing the animals to balance themselves on the slightest level spot in the rocks.

Chamois live in herds of twenty, thirty, sometimes even a hundred. Each herd is led by one experienced female. Before dawn, she leaves her night shelter in the rocky face of the high mountains and, following the same route each time, takes the herd down to the mountain pastures at the timber line. When the sun rises, the herd gradually begins its return to the heights, stopping for rest periods toward the end of the morning in the shade of the bushes. In case of danger, there is an alarm signal of a high-pitched whistle or a stamping of the front paws, and the animals seem to fly away: a chamois can cover twenty-three to forty feet in one bound, ten to thirteen feet in the air. The leader spearheads the flight, followed by the she-goats and baby goats, while the young male goats act as a rear guard. If the leader is hurt, another senior female takes over the leadership.

The males, which ordinarily cherish their independence, rejoin the herd during the mating season, from the beginning of November to mid-December. The male has a gland located behind his horns that manufactures a brownish-green secretion, a sort of musk with a powerful odor. Each male rubs it on the trees, bushes, and grass in one area, defining his territory in this way. No other male is allowed in this territory. The fawn is born after a maximum of 180 days of gestation. At ten days it can follow its mother and nibbles on plants in order to sharpen its teeth. However, its principal food in the first three months is its mother's milk. Once it is weaned it joins the other young goats in a sort of nursery run by a nonbreeding she-goat, and the mother is free once more to graze in peace.

Ibex rest during the day, sheltered from dangers and the wind in some hole in an almost inaccessible rockface. They graze at night, while chamois follows the reverse pattern. ▶

The ibex, which is quite similar to the chamois, is the most common and the noblest of the wild goats of the Alps. It is larger than the chamois: five to five and a half feet long and forty inches at the withers, and it weighs 175 pounds. Its daily

round is exactly the opposite to that of the chamois: it descends to the mountain pastures in the afternoon or at dusk, and returns to its high home at dawn.

Peaceful and shy, the ibex's only defense is flight. It runs as fast as the wind, and no natural obstacle can stop it. But even this speed has been unable to protect the ibex from the threat of extermination: man long hunted it for its organs and its blood, which he used to concoct miracle remedies—especially using its bezoars, which are masticated balls of fur and resin from the stomach of the ibex. Today, more advanced medicines are in vogue, but the ibex is still hunted for personal glory. Whoever can approach, track, and take aim at an ibex is considered a first-rate hunter.

The agile, surefooted chamois climbs the steepest rock faces. When it flees, it runs so lightly and rapidly that the rocks dislodged by its hooves begin to roll away only after the animal has ◀ passed.

Under another name, the famous ibex, the most prestigious of the mountain animals, is one of the constellations of the zodiac: Capricorn. ▶

Lemmings

"So Hans picked up his flute and began to play. The thousands of rats that were devastating the town gathered behind him. He led them to the river, where they all drowned." The old German tale of the Pied Piper has a real counterpart in the suicide march of the lemmings in times of overpopulation.

Even under normal circumstances, the great fecundity of the lemmings is a phenomenon: each female gives birth to five or six babies several times a year. But in some years she is even more prolific. Some scientists have postulated that this fertility results from consumption of vitamin E, which may be stimulated in the plants lemmings eat by favorable climatic conditions. The colony becomes too large to survive, and a certain number of its members seem to be seized by a sort of madness. One dark day a teeming live stream of millions of lemmings begins its march east toward the sea. Wolves, foxes, and stoats follow the apocalyptic procession, devouring all they can, but the army is still not decimated. Eagles, owls, buzzards, and crows hover above, avid for such easy prey. The army of lemmings, racked by hunger and devastated by epidemics, moves relentlessly on, trampling everything in its path, scourging the plants like Attila's hordes. After 125, even 160, miles, the lemmings reach their destination: the icy waters of a northern sea. They have reached the end of their voyage, and they leap into the waves to perish.

In 1518, a German bishop, Olaus Magnus, described one of these bizarre migrations, which are still common in Scandinavia. The mystery has still not been completely unraveled. Three contradictory explanations are offered. According to some, the death march is a degeneration of the collective instinct; to back up this hypothesis, proponents cite the example of rats that devour each other when they become too numerous for their territory. The great geologist Wegener, famous for his theory of continental drift, offers geographic reasons: put simply, Scandinavia was once separated from Greenland only by a narrow channel, and lemmings would swim across this channel in periods of food scarcity to feed on the prairies of Greenland, which Wegener assumed to be their original home. According to Wegener, lemmings still have an ancestral instinct rooted in their behavior, and today in time of hunger still attempt to migrate to Greenland in search of food. A third theory connects the collective suicide of the lemmings to the presence in their blood (as well as in the blood of some other arctic animals) of a kind of "antifreeze." It is thought that this substance, which has been isolated by two American scientists, D. A. Mullen and W. B. Quay, allows animals that possess it to remain active in winter and to survive even the coldest weather, while small mammals that do not have it need to hibernate to resist the cold. According to this third theory, this "antifreeze" sometimes attacks the nervous system and triggers a sort of madness responsible for the mass suicide—which helps control the population.

Aside from their suicidal mania, lemmings which are burrowing rodents, are quite ordinary animals. They are near the hamster in size (five inches) and color (their fur is yellowish brown, with darker areas on the back and lighter ones on the belly, tail, and paws). About a dozen species and subspecies are scattered from Scandinavia to Siberia, in Kamchatka and Alaska, from Hudson's Bay to Greenland. They dig corridors and install their round nests, which they leave only at dusk, in the tangled bush of low plants or just under the surface of the ground. In winter, they build air shafts to their nests through the thick snow. Their life is unvarying, seemingly neither pleasant nor unpleasant—until one day they leave in millions for the sea and death.

The lemming (*Lemmus*), the rodent of the far north, does not hibernate. Perhaps because of a sort of "antifreeze" in its blood, it resists the coldest weather with no appreciable diminishing of its level of activity. This aggressive thick-furred member of the rodent order feeds on lichens and mosses.

The migratory lemmings are doomed. Every so often they leave their overpopulated colonies by the millions, traveling distances of many miles to throw themselves into the sea. There are several theories that attempt to explain this strange behavior, but none has yet been totally accepted.

6. Oak and Beech Forests

The Dormouse

Who knows how bad reputations grow? In the animal kingdom of the Old World, the sweet round-eyed dormouse seems to carry the weight of all the cardinal sins. It is the symbol of laziness—as if sleeping during the day were really so reprehensible for a nocturnal animal, as if hibernating for six months when threatened by death in winter were really a fault. It has a reputation for irritability that leads it to devour its own kind, apparently bothered by their proximity, and yet scientific observation has shown that the dormouse enjoys night excursions with its relatives and will often nest together with others during hibernation. It has been accused of gluttony and petty theft in orchards, and yet who would not take a bite of the young green fruit if there were no ripe fruit to be had?

All the gossip may come from the dormouse's discretion. It lives near man, sometimes even in his barns and granaries, but always manages to remain unseen. It

The dormouse (*Glis glis*) climbs and jumps high in the trees. The young do the same very early. During their first sixteen days of life the father is forbidden access to their nursery.

spends the day asleep. When the sun goes down, one can see its long-whiskered muzzle emerge from a hole, or sometimes from an abandoned squirrel or large bird nest. It will glance hurriedly to left and right; then its body follows, then its bushy tail—almost as long as the whole body, six to seven inches. Guided by its senses of taste, smell, and especially touch, it sets out to find vegetable food: acorns, beechnuts, walnuts, chestnuts, and fruit and grapes—a preference that is naturally not well received by orchard and vineyard owners.

Dormice live in southern, central, and eastern Europe, and in some parts of the Near East. They like forests, nimbly climbing trees by planting their claws in the bark; they dive like squirrels, using their tails as a rudder.

After the grape harvest, they settle down for the winter, sometimes in a burrow a yard deep in the ground, or in an old building, the hollow of a tree, or a hole in a rock. They build a soft nest, tuck the tail around the body as a comforter, and sink into a near coma: their heartbeat slows down, their body temperature drops drastically, their limbs become rigid, and they may lose up to 50 percent of their body weight.

The details of the mating habits and family life of the dormouse have been observed in dormice in captivity. They mate in May, just after awaking from hibernation. The female then immediately isolates herself and prepares "delivery" quarters in a nest or tree hollow. After a maximum of six weeks, she gives birth to a litter of three to six bald, pink, blind baby dormice. She nurses them until they can see, about three weeks. The father stays away until their fur has grown; he waits patiently for about sixteen days, then rushes in to meet his family, which he grooms. From then on he ferociously defends the litter against intruders. As soon as the babies can climb, they follow their father to his nest, returning to the mother when they are hungry. Often father, mother, and babies will spend the winter asleep together, separating permanently in the spring, during the mating season.

▲ When the acorns and beechnuts become scarce, dormice munch on ripe fruit in orchards. The growers might forgive even this if the dormice didn't bite into the green fruit as well.

▲
Eyes shaped like two black marbles and a long tactile mustache indicate that the dormouse is a nocturnal animal.

A bushy tail has many uses: it functions as a rudder for jumping, and in winter it becomes a comforter under which the edible dormouse hibernates. ▶

A comic little pink package, a baby dormouse that has just run off for the first time is carefully and firmly returned to the nest by its mother. ▶

The Long-Eared Owl

Armed with cruel feathered talons, the long-eared ▶ owl feeds on field mice. Here it carries its prey to its eyrie and offers it to the female, who will tear it into small pieces to feed to her young.

When some animals feel threatened they hide their heads. The long-eared owl doesn't stoop to such low tactics and disdains even any indication of fleeing. In its spotted and striped yellowish-brown outfit, it is sure of its camouflage—indeed, so sure of melting into the surrounding branches that it does not fly away when approached.

The long-eared owl is found all over Europe as well as in North Africa and North America. With no apparent preference for one spot over another, it frequents the seedbeds of cone-bearing trees, forest foliage, isolated small woods, parkland—and cemeteries, where it is the symbol of bad luck. It sleeps during the day, perched upright on a branch, its characteristic two straight brown feathers on its head sticking up above its ears. It hunts at night, relying on its piercing eyes and fine ears; it grasps its victims in its feather-covered talons, kills them with one blow on the nape of the neck, and swallows them with no more ado. Bone, claws, teeth, fur, and feathers cannot be digested, and they are regurgitated in wads. Examination of these provides some idea of the voracity of this bird: the balls regurgitated by seventeen long-eared owls over a period of two and a half months revealed the remains of 1,528 field mice, their favorite prey, representing 92 percent of the total number of animals consumed.

The long-eared owl has one important advantage in its night hunting: because of its fluffy, ventilated feathers, laced with an anterior row of quills, it flies completely silently.

In the autumn, when field mice disappear for their winter's sleep, groups of long-eared owls will often assemble and fly south, since their prey animals there do not hibernate. In the absence of field mice, those individuals that do not migrate declare war on sparrows, finches, and green linnets.

These predators are squatters, and therefore never build nests; they settle down wherever condors, crows, pigeons, or squirrels have left vacant homes. In early spring, the female lays four to seven eggs, at intervals of two to eight days, which she immediately broods. The parents feed the young small bits of meat, but only for a short period of time. Then, just like the adults, the young must catch their own field mice.

▲ All covered with white down, young long-eared owls (*Asio otus*) are born in a nest left by another bird or a squirrel.

The European Stag Beetle

The earth is overrun with shiny-carapaced beetles. Entomologists have counted more than 250,000 different species with intriguing names: bombardier beetles, whirligig beetles, carrion beetles, click beetles, and so on. The stag beetle family (*Lucanidae*) alone is composed of 1,100 different species.

The stag beetle is the largest beetle in Europe; the male can be three inches long. Its most striking characteristic is its warlike—and ridiculous—appearance: the more highly developed it is, the more its antler-like jaws gets in its way. In duels fought over females, the male with the shorter mandibles will win, as it has the harder bite. The shorter females have only nippers, but often these are prominent and very strong.

The stag beetle needs only two trees in order to feed and shelter itself, reproduce, and survive: the chestnut or the oak. The female lays her eggs in the worm-eaten trunk of the tree. The larvae remain there for five years, feeding on the rotting wood. Then the larva becomes a chrysalis, out of which a year later emerges a stag beetle. It is already adult-sized, like all insects that go through a complete metamorphosis. Finally, it is from the sap of the live tree that the stag beetle nourishes itself in its final stage, during the summer; at the end of the summer, it dies.

New observers are always surprised by the confidence with which the beetle

The stag beetle has a brush-shaped lower "lip" ▶ that it uses to lick off the sap of a "bleeding" oak tree. It moves on when its source of nourishment is depleted.

Right:
In spite of its fierce appearance, the shiny brownish-black stag beetle (*Lucanus cervus*), the largest European beetle, is totally harmless. Unfortunately, this beetle is threatened with extinction.

moves, at night as well as during the day. Stag beetles are equipped with a rather special sensory system: the ends of their antennae spread out into "leaves," which are longer and thicker in the male than in the female; these structures have tiny olfactory holes. When the stag beetle walks or flies, it spreads its antennae like a fan, far from its body; guided in this way by odors, it manages not to bump into things.

In some countries, such as Germany, Austria, and Switzerland, the stag beetle is protected. The passion of collectors and, more important, the chopping down of the old oaks (which have become increasingly rare in the developed countries), are threatening the stag beetle with total extinction in Europe.

Above:
The oak or the chestnut is indispensable for the stag beetle, as it is the source of food and life support. The female lays her eggs in worm-eaten tree trunks, where the larvae are born, feed, and grow. The sap from live trees is the adult's sole diet. The complete metamorphosis of this insect, from an egg to a stag beetle, passing through the stages of larva and chrysalis, lasts at least six years.

More for display than for war, the "antlers" of the male stag beetle, which impress the adversary and attract the females, are cumbersome for combat. ▶

◀ The leaf-shaped antennae of the stag beetle (the female is shown here) have olfactory organs. These allow the beetle to smell and thus orient itself.

Blackbirds, Song Thrushes, Orioles

It is dawn in the city: trucks, subways, sirens, smog . . . the sounds and sights of daily urban life are starting again. But, should a garden, a square, a park—perhaps only three trees and a few patches of grass—have managed to survive somewhere between two buildings, then suddenly a miracle occurs: the joyful song of a bird breaks out in glorious trills toward the rising sun. The blackbird greets the city dweller, who gladly hears the song without understanding why this forest dweller now comes closer and closer to man, moving into the heart of his cities.

There are blackbirds all over Europe, as far as mid-Scandinavia, as well as in western Asia and North Africa. Once they move into an area, they behave like local toughs—slightly cheeky, quarrelsome, brawling, but nonetheless effective: their loud whistle when they are threatened by danger is an alarm signal for the whole cheeping band, and frightens off egg thieves and nest pilferers. When there is no danger, blackbirds hop about on the ground, balancing on their tails, endlessly searching the grass for spiders, larvae, beetles, and earthworms, which they root out with their chisel-like beaks. Along with starlings, they celebrate the cherry season with feasts in the orchards. Clément, the famous French songwriter, wrote: "When we sing of the cherry season, we must sing of the mocking blackbird too." Scarecrows have never really succeeded in stopping these picnics. And the cherry season is followed by the ripening of various berries and wild roses.

The blackbird is a remarkable nest builder: it skillfully interlaces moss and dry straw, cements the structure with mud, and lines it with feathers and soft grass. This bowl-shaped shelter can survive two or three nestfuls of birds. Each is made up of four or five carefully brooded eggs; the female never leaves her post for more than the brief time necessary to eat during the hot midday. The baby blackbirds, born blind and bald, have an insatiable appetite that keeps the parents busy at all times. The slightest vibration of the nest makes the beaks of the baby birds open up and cry out for food: the gaping red gullets never stop demanding the caterpillars, insects, and earthworms that the baby birds need to grow.

The calls of the blackbird and the song thrush are distinguished by their musical sophistication: while the blackbird is constantly improvising new tunes, the song thrush sticks to one basic theme varied in two or three different ways. It inhabits western, central, and northern Europe, at all altitudes, and, like the blackbirds, en-

◀ The gaping gullet of the baby bird endlessly announces its hunger.

An equally fierce hunger tortures the small song thrushes (*Turdus philomelos*). It is best for the young to take advantage of their first sweet days: once they leave the nest, childhood is over and they never return. ▶

The father blackbird (*Turdus merula*) is constantly ▲ occupied finding caterpillars and insects to feed his family.

"I value my garden more for being full of blackbirds than of cherries, and very frankly give them fruit for their songs." *Joseph Addison.* ▼

▲
Just born, the golden oriole's young *(Oriolus oriolus)* clutch the edge of their swaying nest to avoid falling out.

joys the proximity of man. It adds one special dish to the usual fruit and insect fare: snails, which they find under mucus-coated rocks near the nests. The thrush hits the shell against the rocks and easily extracts the animal. Among blackbirds, only the young migrate south in winter, but all song thrushes fly to the Mediterranean countries come October.

A third bird delights the ear of the nature lover: the golden oriole, with its crystalline arpeggios. It is a more distant relative of the blackbird, and like it a member of the order *Passeriformes*. But its bright yellow feathers are seen quite rarely, since the oriole prefers the tree tops. Over most of Europe, the male and female build their nest together, out of grass, climbing plants, and saliva-coated roots. Hanging from the fork or some oak or birch, the nest sways in the slightest wind, and one can see the three to five baby birds of an annual nesting clutching tightly to the edge. At about fifteen days, the young orioles make their first test flight, and their first attempts to sing. Then, after migration, they have the entire winter to practice under the African sky.

Just able to fly, the young cannot yet feed themselves. They sit on a branch and the feeding ritual begins.

A little impudent and provocative, the blackbird seems to be constantly on the lookout. It uses its eyes to observe its surroundings, and at the slightest alarm flies off whistling so loudly that its fellows—and the entire neighborhood—are ▼ alerted.

The Red Fox

King Solomon may have given the fox its bad name when in the Song of Songs he wrote of "the little foxes that spoil the vines." No matter how many explications have followed since, pointing out that the foxes in the Bible are not those we know, and no matter how many scholars have tried to prove that only fairy-tale foxes like grapes, the popular image remains. The red fox is a thief, people say. The red fox has no respect for the rights of others; nothing can inhibit it from raiding chicken coops. Forest rangers may protest, "What is one chicken happily stolen, compared to the harmful animals such as rats, mice, moles, and snakes killed by the red fox, not to mention the carcasses it disposes of?" But even this has no effect. In fact, ill will toward the red fox has increased.

Red foxes are found in all the forests of Eurasia, in Scandinavia and the Himalayas, in North America and Africa north of the Sahara, classified into nearly fifty subspecies. They are sometimes red and

Left:
Young red foxes undertake their first hunting expeditions in teams. In the autumn, the family groups disband and each individual then lives for itself.

Right:
The ears of this handsome red fox (*Vulpes vulpes*) are pricked up and its pupils gleam—their "phosphorescence" is due to refracting cells located at the back of the eye.

sometimes brown or yellow, always slender, lithe, large-eyed, fine-muzzled, and bushy-tailed. They are remarkable builders: each burrow is constructed according to a functional plan, with branching corridors, entrances, exits, and even emergency exits. They are occupied by successive generations, since red foxes are creatures of habit and dislike moving.

Once a year, around February, the usually quiet red fox becomes a yelper—it is the mating season. The males don't go so far as to fight over a female, but they do quarrel loudly among themselves. Two months later, in the spring, the litters of four to six young are born, as fat as moles and as gray as mice. At four weeks they gambol and frolic in front of their lair. Father and mother supply and care for the offspring, but even together barely manage to satisfy the hunger of the voracious little red foxes. Sometimes a female with young finds herself without a mate and, the ends justifying the means, she commits the crime of raiding a chicken coop. Little by little, the young learn to live alone. In autumn, the family disbands.

Like many wild animals in temperate zones, red foxes, who do not hibernate, suffer from cold and hunger. Throughout the winter they track hares and roe deer over the frozen or snowy ground; sometimes they even attack weakened adult roe deer. It goes without saying that they are blamed for this ignoble behavior. But their friends invoke the law of the jungle.

The Roe Deer

Like its graceful silhouette, its shiny coat, and its mournful eyes, the roe deer should only be described in the lightest, most delicate, of words. In its first year, the roe deer grows two bony knobs on its forehead, the beginning of antler development. These grow into spikes, and the animal is now called a brocket. The spikes, still soft, are protected and nourished by a sheath of skin known as velvet. In April, the brocket sheds this velvet by rubbing up against branches. The antlers harden, then become brittle and fall off. New spikes, with tines, grow in, and once again, the following year, there is the shedding of velvet in the spring, the loss of antlers in the fall, and the growth of new antlers with yet more tines. Toward the end of the winter, the roe deer gets its "second head." The cycle repeats itself unalteringly. The older the male the thicker his antlers will grow in, and the heavier—well over a pound—and longer (twelve inches) they will be, their larger base decorated with small pearl-shaped outgrowths of bone.

In its social life, the adult roe deer practices monogamy—and still keeps a harem. At mating time, the male imperiously claims a female and acts very jealously to-

▲ Wearing a shiny yellowish-brown or reddish-brown coat spotted with white, the immobile fawn blends into the light-streaked underbrush.

Above left:
The lovely, fragile fawns wait for the May sun before they come into the world.

◀ The cautious roe deer is never far from her young. Man must never touch a fawn, even if it appears to have been abandoned; the mother may reject a fawn that bears an alien smell.

ward her. But once mating has occurred, he sends her back to the "female quarters," a group composed of several females and their young which are still dependent on them. There the young fawns are born. As fertilization occurred in July or August, the young should logically be born in the winter, when they would be endangered by hunger, the cold, and the more than usually hungry carnivores of the forest. But the fawn is cleverly protected by nature: the implantation of the fertilized ovum in the uterine wall is postponed, and the embryo only begins to develop in December. Thus the fawn is born in May, nine and a half months after conception, when the leaves can provide shelter and its enemies are less hungry. In the following months, the mother reverses her life rhythm: while she used to graze at

dusk, she now becomes diurnal, knowing that as long as there is light she can graze in peace while her fawn is in no danger. She is never far away from her fawn and will defend it—but if one finds a fawn by itself, apparently lost in the underbrush, one mustn't touch it: the mother is likely to ignore her offspring completely if she detects an unfamiliar odor on it.

The roe deer, the smallest and most common of the antlered animals of the Old World, inhabits all the forests of Europe, as far as Asia. It is sometimes found in the heart of the forest, sometimes on its edge, and sometimes even in small woods. It is forty-four to fifty-two inches long and stands thirty inches at the withers. At less than an inch, its tail seems ridiculously short. This lack is compensated for by a large, shiny white spot at its base. At dusk this white patch is a sort of "rear light" that allows the fawn to follow its mother.

Each herd has its own territory and standard itinerary. Roe deer feed on buds, tender leaves, and fine grass; they also like farm produce, lettuce and young cereal shoots, and they can devastate sown fields.

Winter is the season of famine and death for the roe deer. When desperate, they will graze on anything they find, frozen grass and stunted lichen, and they will even sometimes invade private gardens. Forest rangers often set out food for them, but often they cannot all reach it, and thus many die long before attaining their maximum life span of seventeen years.

The white patch at the base of the tail allows the fawns to follow their mothers easily. ▶

◀ At dusk, the herd seeks its grazing areas.

The tines of the roe deer (*Capreolus capreolus*) are sheathed in velvet. The deer sheds the velvet in the spring and has bare antlers until autumn, as shown here. ▼

◀ Winter brings snow and distress. Often, the exhausted, famished animals, threatened by carnivores that are hungry too, cannot reach the food left for them by rangers at the edge of the forest.

In times of scarcity roe deer will eat dried-up plants, tree bark, and stunted grass. In spite of its shyness, the roe deer will even raid planted gardens. But the winter goes on, and the weakest die.

The Common Toad

One must not confuse old wives' tales with fairy tales. In the first, the toad is characterized as evil, as the helpmate of the witch and responsible for all sorts of abominations. But fairy tales tell us not to judge by appearances alone—because here the toad is often part of some magical test. When the hero, or more often the heroine, shows the least interest or pity for this creature, all sorts of wonderful gifts are showered down. Gardeners tend to agree with this more favorable interpretation—they are very fond of these leaping creatures that frolic about in their lettuce beds and rid them of slugs, larvae,

◀ Though ugly, the common toad (*Bufo bufo*) has lovely golden eyes. It is welcome in gardens, as it kills slugs and destructive insects.

This batrachian poetically sings its love song in the moonlight. Guided by its infallible sense of direction, it finds its spawning grounds in pools and ponds. ▶ ▼

insects, and various garden pests.

The common toad inhabits all of Europe, North Africa, and the temperate zone of Asia as far as Japan. They acclimate as easily to plains as to mountains as high as 6,500 feet, to humid areas, gardens, meadows, and forests. During the day they are discreet and remain hidden under some rock, tuft of grass, or in a hole. At night they hunt for food in small, clumsy leaps, their lasso-like tongue projected forward. When a toad catches something large—a beetle, for example—it uses its toes to help push the prey into its mouth. Quite reasonably, the toad will stop eating when it is full and return home, guided by its keen sense of direction. But its moderate appetite does not prevent it from growing fat in summer when food supplies are especially plentiful. It then becomes uncomfortable in its skin, which does not grow with it, and sheds it. Shedding takes place at three- to ten-day intervals.

The toad is endowed by nature with two means of defense against its natural enemies: glands, located all over its body, that secrete a noxious liquid; and the ability to stand up and swell itself up to frighten off hedgehogs and snakes (although these are not always successfully fooled).

Like the song of the birds, the toad's piping announces the spring. In March evenings, the toad has returned to its spawning grounds, which are often in distant pools or ponds, it vocalizes in earnest. This is the call to the female: the eggs are to be fertilized now. After coupling, the toad drapes the strands of spawn, ten to sixteen feet long and composed of thousands of black eggs, over rocks in the water or around the stems of water lilies and other aquatic plants. The sun does the rest. The gelatinous envelope around the eggs

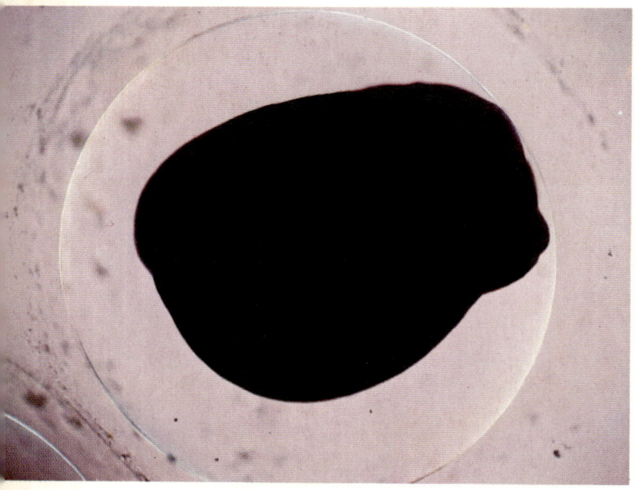

◀ The eggs, which are at first round, soon take on the shape of a heavy comma: this is the beginning of the development of the tadpole larva.

The threads of spawn, ten to sixteen feet long, are carefully wound around rocks and aquatic plants. ▼

then dissolves, and incubation occurs in the heat. Soon the tiny black dots become thinner at the tail; by the end of three months they have developed into tadpoles. Bundles of external gills allow the tadpoles to breathe in the water. They feed on plants and propel themselves by rowing with their tail. The hind legs grow first, and then the front legs. The oarlike tail shortens, the gills disappear and lungs develop internally. The small toad then moves onto dry land.

During the winter the immobile toad hibernates in a leaf-filled hole. It is no longer breathing: its skin provides the necessary exchange of oxygen. A frost below 32° F. is fatal. The toad is capable of reproduction after the age of five. Given mild weather, it may live thirty or even forty years more—providing it is not devoured by its natural enemies (or killed by man).

An oarlike tail allows the tadpole to swim very rapidly; here its hind legs are starting to develop. Once it has consumed the contents of its nutritive yolky pouch, it feeds on enormous quantities of algae and small organisms. ▶

The complete metamorphosis: the tadpole is now a small toad; it climbs onto the ground and finds a place in the sun. ▶

The next to the last stage: the hind legs, then the front legs, have appeared, and the tail has shortened.
▼

The European Badger

When a badger is spotted, it is quickly tracked and hunted. But this rarely happens: most of the time this rather grumpy creature avoids man, other animals, and even its own species. Yet the badger actually may be the most peaceable and least harmful of all the European carnivores.

It has a special liking for comfort. A good burrower, it is careful to dig its hole behind nettle thickets on the sunny slope of a hill or under an isolated farm; thus central heating is guaranteed. There are four to eight entrances to the badger's main dwelling. The central chamber is equipped with vertical air vents; several corridors lead to the vast moss-filled "bedroom," which is always kept meticulously clean. The fussy and unsharing badger allows no intruders into its home. If another badger tries to get in, there is furious combat, and if a fox attempts to enter, the badger becomes enraged.

Badgers, which live alone, are thirty inches long and twelve inches high, and have elongated heads. They have the same life style in all the wooded areas of Europe and Asia. They rest during the day and seek food at night. They are not very particular and will eat anything they happen to find: acorns, beechnuts, berries, mushrooms, fallen fruit, insects, beetles, snails, grasshoppers, mice. They dig up roots as well and lick off wild beehives. By autumn they have become so fat that their belly almost drags the ground. This reserve of fat allows them to survive a sleepy winter in the cozy warmth of their home, whose entrances they stop up with wads of leaves. Their hibernation is neither very deep nor very long. Once the weather turns milder, the badger thrusts out its nose, sniffs the air, goes for a drink, trots about, sees to its affairs, and then goes home again.

Twice a year, males and females put aside their disputes and meet. After mat-

Far left:
The favorite foods of the American badger (*Taxidea taxus*) are ground squirrels and prairie dogs.

◀ At four months, the clumsy young European badgers leave their home for the first time to play in the sun. Their mother raises them by herself.

▲
◀ The clever European badger (*Meles meles*) equips its hole with several well-camouflaged holes that are often as steep as toboggan runs.

ing, they even engage in a sort of honeymoon. Like that of the roe deer, the first litter, the result of summer mating, has a delayed development that allows the young to be born at a favorable time and in a comfortable nest; the second litter, from the November mating, has a gestation period of only thirty-five days. The female cares for the young, which are as tiny as mice and covered with a white woolly fur, by herself. She nurses them for three months, then feeds them regurgitated food. At four months they play in front of their burrow in the sun and gradually learn everything a badger must know to survive independently. Once the apprenticeship is over—including field work in night hunting—the mother brusquely throws the young out. Then they move out into the forest and, being badgers, the first thing they do is build themselves comfortable new homes.

The Koala

With its round head, innocent ears, and button eyes, its sweet bulbous nose and its soft fur, the twenty-four-inch koala certainly looks like our childhood Teddy bear come to life. And though we may be forced to admit that the koala is not a bear at all—it is one of the climbing marsupials, inhabiting Australia like its cousin, the kangaroo—it doesn't really matter: the image of the Teddy bear is strong for us.

This living toy likes only trees, and any tree at that. It is particularly fond of the treetops of the fresh-smelling eucalyptus and feeds solely on its leaves. As each variety of eucalyptus tree secretes different poisonous substances at different times, the koala proceeds from one kind to another, according to the season. The leaves provide both food and water, and the koala never drinks; indeed, the word "koala" means "he who doesn't drink."

The koala is ideally suited for life in the branches, where it even spends its nights wedged in a fork. Its front paws, with their opposable thumbs, are well formed for grabbing, and its hind feet have a big toe that separates from the others, allowing the koala to perform daring acrobatics. Once a year, the female gives birth to one, or more rarely two, little koalas, which are less than an inch long at birth. The baby stays in the mother's pouch for six months. From the first month, the mother feeds it a gruel of regurgitated eucalyptus leaves. At six or seven months, when it is six inches long, the baby climbs on the mother's back and is carried around this way for a year.

The prototype of the Teddy bear, the koala (*Phascolarctos cinereus*) is gay and lively, as well as downright lazy: after leaving its mother's pouch, the baby koala spends another year carried on her back.

7. Open Woods and Scrubland

The Barn Owl

▲
The pale heart-shaped face of the barn owl (*Tyto alba*) is in fact a mask of feathers. Its pensive look makes it appear to be dreaming while it sleeps
◀ during the day in its shelter, eyes wide open.

Night is the barn owl's kingdom. Then it is a hunter, showing a marked preference for field mice. But if there is a lack of rodents, small birds will pay the price. ▶

Of all owls, probably the most familiar is the barn owl, with its pensive, sphinxlike face. It always settles close to man, in a barn or granary, a nook or a hole in a high wall—when it doesn't choose the village belfry, that is. In return for shelter, it undertakes the destruction of whatever animal pests the area offers in the way of field mice (its special treat), shrews, an occasional mole, and rats. Sometimes, however, it adds a few small birds to its diet: sparrows, swallows, swifts, and redstarts.

Divided into many subspecies, barn owls live almost everywhere throughout the world except in the far north. Especially well adapted for night hunting, their eyesight can pick up the most fleeting glimmer, and their hearing is sensitive to the slightest rustling—great advantages on dark nights. Like all owls, they fly quite

Claws at the ready, the barn owl silently approaches the mouse, which is unconscious of its danger. The bird's immense eyes enable it to perceive the most fleeting light that reveals the presence of a prey. On dark nights, the barn owl is infallibly guided by its very sensitive hearing. ▶

soundlessly, swaying slightly, like bats. Their victims rarely hear them approach: seized by the owl's claws, they are killed by a single blow of the hooked beak and immediately swallowed whole. The remains are regurgitated in pellets: examination has shown that 74 percent of the barn owl's nourishment comes from field mice and 10.5 percent from shrews. Because of its pest-control services, the barn owl, like most owls, is protected in many parts of the world.

This nocturnal bird, easily identifiable by its knock-kneed legs, likes to spend the day sitting on a branch, taking the sun.

In early spring, it sounds its mating call: rumbling, loud rasping sounds, sad wailing, and hooting. Without bothering to build a nest, the female lays four to six eggs, and occasionally ten or eleven, sitting on them immediately, even if she hasn't finished laying them all. The young are hatched thirty to thirty-four days later, all white and grimacing like monkeys. They take their first flight between seven and nine weeks later.

The male brings the food—as many as ten mice for one feeding—during the period that the female is hatching the eggs. Then both parents share in providing food for the brood. Father and mother vow long fidelity: barn owl couples have been observed still together after eight years of conjugal life.

The Hedgehog

At dusk, the hedgehog (*Erinaceus europaeus*) ▶ leaves its nest to search for food.

A ball of needles mounted on four legs, the hedgehog is protected by an armor of 16,000 spines. When all is well, these spines lie flat, but when danger threatens, a circular muscle sheath contracts, raising the quills, and the hedgehog becomes as spiny as a cactus and as impregnable as a fortress. It is so confident of its defenses that it has not learned to avoid roads, and thus its greatest enemy today is the automobile.

The hedgehog is one of the most effective destroyers of animal pests, as its prey are worms, mice, snails, and snakes. It dozes during the day and hunts for food at night. It is as comfortable in the fields as in the woods, in gardens and in parks, and will even approach man. Eleven inches long, six inches high, with a small inch-long tail, a pointed muzzle and bright shoebutton eyes, the hedgehog is found all over Europe and in the temperate areas of Asia and the Middle East.

As soon as the first signs of winter appear, the hedgehog prepares its special quarters: a hole barely a foot underground or a nook in a pile of twigs or leaves. The hedgehog furnishes its nest with leaves, grass, and moss; at the first frost it retreats to this cozy shelter, rolls up into a ball, and falls into a deep state of lethargy. Its body temperature drops to 43° F., its heartbeat falls to 20 beats a minute (as contrasted to the normal 181). About once a month, or more frequently if its body temperature drops too severely, it rouses itself, goes out for a short walk, trots about to warm up, then returns to its nest and falls back into its stupor. In March, the hedgehog becomes active again.

Hedgehogs live a solitary life except during the mating season. After a gestation period of five or six weeks, the mother gives birth to four to seven tiny, thumb-sized babies, in the warmth of a tight nest. At birth, the quills of the young are soft; the skin underneath, which is as thick as upholstery, is also soft, and hardens very gradually. Thus all are safe from injury in the close quarters of the nest. The mother nurses her young for seven weeks, and then feeds them worms and insects. They soon accompany her on her night excursions to learn how to catch food and generally how to fend for themselves. They have one difficult period to survive: their first winter. Many inexperienced young hedgehogs do not bury themselves deeply enough in their burrows and die of the cold.

▲ During their first three weeks the tiny, blind hedgehogs depend totally on their mother. Their quills are still soft.

◀ This armored creature sleeps during the day in its shelter.

In case of danger, the contraction of a circular muscle sheath under the skin raises the coat of spines. Head, tail, feet, and belly seem buried in the mass of quills. ▶

Hawk Moths

If all the Lepidoptera—that is, the moths and butterflies—in the world were to fly wing to wing, a shimmering roof would appear over our heads; there are more than 150,000 different species, and the characteristics of each one could itself fill a book. Thus it is more useful to examine their common characteristics: wings covered with microscopic, multicolored scales; compound eyes made up of thousands of facets, 4,000 to 30,000 depending on the species, sensitive antennae (that only in butterflies, not moths, have enlarged, knoblike tips) with olfactory organs; and most important, some moths and some butterflies have an extraordinary coloration that allows the defenseless caterpillar to fight off attackers with its gaudy minimonster aspect, whereas the adult conceals its bright beauty by folding back its wings to expose their duller underside and thus blends in with the nearest object, a dead leaf, a fallen petal, a pebble, a piece of bark.

The scales on the wings develop in the cocoon from fine hairlike processes. These become flattened out, puffed with insulating air, and piled in overlapping layers like the tiles on a roof by the time the moth—or butterfly—is ready to emerge. When the wings are touched, these scales come off on the fingers as a

The marvelous cycle of metamorphosis has begun. From the egg of the peacock hawk moth (*Eudia pavonia*), a pretty multicolored caterpillar has emerged. It will have one sole occupation: feeding itself. Set off by some mysterious internal mechanism, one day it will begin to spin its cocoon, and will spend the entire winter in this silken covering. Gradually, the moth is developing inside. In April it will emerge from the cocoon, spread its wings, and fly away.

bright powder. Removal of the scales can be fatal, since the now-bared wings allow the air to pass through, making it impossible for this insect to fly and survive.

The olfactory organs on the moth's antennae allow them to move unerringly in the dark toward the fragrant flowers on which they feed. In some moths, the sense of hearing is fixed on the wavelength of the bat's cry, warning them of the approach of that predator in time for them to seek shelter.

One of the best known groups of moths are the hawk moths, which are distributed throughout the world. The body, which is compact and covered with fine hairs, is spindle-shaped; there are two short variegated back wings and two long, narrow, elegantly streamlined front wings that fold up into a roof over the body and the back wings when the moth alights; then the moth blends absolutely into the branch or leaf on which it is perched. Many hawk moths fly and feed during the day, stopping at flowers, where they hover and vibrate in place, like hummingbirds. More than a thousand different varieties of hawk moths have been classified. Most are exotic; an Australian species has a wingspan of eight to nine inches. There are about a hundred species in North America, and twenty-two species are found in Europe, including

Is this a frightening monster, a fire-breathing dragon? No—it is only a harmless caterpillar, whose appearance is its only, passive, form of defense. Shown here is the caterpillar of the vine hawk moth (*Dellephila elpenor*); at right, that of the little vine hawk moth (*Pergesa porcellus*).

the death's head moth, so-called because of the macabre skull-like markings on the thorax. This species has a relatively short proboscis that often cannot reach the heart of flowers, the source of nectar. In search of substitute food, it is frequently led by the sweet perfume of honey into a beehive—and is killed by the stings of the occupants. Its relatives are better provided for: the moro hawk moth, for example, can plunge into the heart of the deepest corolla and extract the nectar. The bindweed hawk moth has a proboscis of five to six inches; that of one American species is eleven inches long. In this species the proboscis rolls up in a spiral and tucks under the moth's head when the moth is at rest.

Like birds, some moths are migratory—but in the opposite direction. In autumn the moro hawk moth leaves the far south for central Europe, where it comes to die. The death's head moth, the oleander hawk moth, and the bindweed hawk moth leave North Africa or southern Europe for the Baltic shores, flying over the Alps. These species winter in Europe, where they lay their eggs; some of the parent generation even survive to make the return trip.

While the adult hawk moth owes its safety to its discretion and its successful camouflage, the caterpillar is protected by a pronounced aggressiveness of markings and size: variegated, speckled, sometimes as long as six inches, it looks like a tiny monster, and thus frightens and disconcerts its enemies.

Feeding solely on the leaves of the spurge, the caterpillar of the spurge hawk moth (*Celerio euphorbiae*) becomes a slender two-and-a-half-inch moth (*right*) that is found all over central and southern Europe. ▲ ▶

One finds the caterpillar of the privet hawk moth (*Sphinx ligustri*) on the edge of meadows and in city gardens as well: it lives in lilac bushes. ▶

The sweetbrier and the potato plant serve indiscriminately as homes for the death's head hawk moth (*Acherontia atropos*); the female is shown here. Each year this moth flies over the Alps on its migration from North Africa to the Baltic.
▼

▲
The sweet-smelling beehive tempts the death's head hawk moth, whose short proboscis cannot extract enough nectar from flowers. This gluttony will cost it its life, for the bees will sting it to death.

The Ring-Necked Pheasant

"This bird is no earthly creature—see its beak, see its plumage," sings Charles Trenet. One is inclined to agree on seeing the pheasant suddenly open its multicolored fan. And indeed, the ring-necked pheasant does have an exotic ancestry, albeit of this world; like its close relative, the peacock, it comes originally from the Far East and southern and central Asia. But it has adapted perfectly in Europe and North America.

The pheasant is a wandering bird, with no fixed home except that prepared for the nesting eggs. It enjoys bushes or under-

◀ The proud ring-necked pheasant (*Phasianus colchidus*) inherited its beautiful multicolored plumage from its Asiatic ancestors. Its train can be as long as thirty-four inches.

brush that conceal it so that it can feed in peace, and is most frequently found near streams and ponds. During the day, it runs along the ground poking incessantly to find insects, larvae, snails, lizards, blindworms, buds, berries, and fruit; at night, with one beating of its wings, which sounds like ripping silk, it propels itself into a tree to sleep. The male can be as long as thirty-six inches, and the female, twenty-six.

All winter, pheasants travel in groups. During the mating season in March, each male establishes a territory and begins his complicated ballet of courtship. There are more than 250 pheasant species and subspecies, and each has its own way of declaring its sentiments, involving a great flurry of wings and display of colors. The golden pheasant and the Lady Amherst pheasant do not face the female, as the other birds of this order (Galliformes) do, but stand at her side and suddenly unfurl a dazzling fan of feathers. The male of one of the species of Borneo has an erectile crest, which when raised greatly changes the bird's entire appearance. Many varieties display sumptuous tail effects, like the silver pheasant, whose colors are iridescent, or the Reeves's pheasant, whose train is almost six feet long.

The ring-necked pheasant hen broods about eight, twelve, or sometimes even twenty eggs, for which she prepares a nest in the shelter of a hedge or bush. After about twelve days in the nest, the young pheasants are capable of fluttering to a nearby tree.

According to some ornithologists, the ring-necked pheasants that are hunted today all over Europe and North America are the result of the crossing of pheasant populations from the western Caucasus, central Asia, China, and Japan. This would explain their great diversity of appearance.

The underprivileged hen must be content with a sober dress of yellowish brown speckled with white, and her tail is much shorter than the male's. She broods her eggs on the ground. Pheasants prefer to wander in areas of high grass and bushes, which provide cover in case of danger.
▼

Gall-Producing Wasps

Left:
These coin shaped oak galls were formed by the gall wasp (*Trigonaspis megaptera*).

◀ These onion-shaped galls on a beech leaf are actually not the work of a gall wasp at all, but of a gall fly which is similar in appearance to a mosquito.

Every enigma in nature is like a tangled ball of string for scientists to try to unravel. No one has yet succeeded in establishing the precise relationship between gall wasps and the plants on which they live parasitically. The effect is easily seen: the gall formed on the plant. But the cause has not yet been understood. Why does the plant so readily accept these guests?

Gall wasps are minuscule creatures, several millimeters long, which are related to the ichneumon wasps. They are shiny black or dark brown, and reproduce in an unusual way. In a number of species, sexual and asexual generations alternate, the asexual generation reproducing by parthenogenesis.

The female uses her ovipositor to pierce whichever plant, or part thereof, her species prefers (often it is the oak), and lays her eggs on its surface. A larva emerges from each egg and soon settles down in its cell and sets to its single task of feeding itself. It does this untiringly and with great ease: one can see outgrowths on the plant corresponding to the needs of the parasite. Is this just a matter of nourishment, or do the galls provide protection as well? Some observers have suggested the hypothesis that secretions of the larva are responsible for the development of the gall, but this theory has not been confirmed.

On the oak, one can examine the development of two successive generations of common or oak leaf gall wasps *(Diplolepis quercus folii)*. This species is identifiable by the galls they form on the underside of oak leaves; these are yellow in spring and red in summer, and somewhat smaller than a dime. The sexual generation produces larvae during the winter, which become chrysalises in the spring. The adult gall wasps that emerge are shiny black and four millimeters long. After mating, the female stings an oak leaf and lays her eggs. The summer generation thus produced is asexual. This generation spends its larval stage in the spongy galls. At the end of autumn, asexual adults emerge, which sting the buds of the new leaves and deposit their unfertilized eggs. From these will come a new, sexual generation; the cycle is complete. The galls produced by this species once had economic value: tannic acid and ink were extracted from them. But today this process is no longer used.

Other oak gall wasps produce small

One stage in the cycle: at the end of autumn, the oak leaf gall wasp (*Diplolepis quercus folii*) of the asexual generation emerges from the gall. It immediately lays unfertilized eggs in a leaf bud, which it pierces with its ovipositor. In spring male and female gall wasps capable of sexual reproduction will be born from these eggs. The second generation females will deposit their fertilized eggs in the veins of the leaves, where new galls will then grow.

Willows, too, are affected by galls. The willow sawfly (*Pontania proxima*) deposits its larvae on the leaves, and these bean-shaped outgrowths appear.

The black alders of Europe are relatively protected from the alder sawfly (*Fenusa dohrni*). In the United States, however, this variety is responsible for great devastation.

Tannic acid and ink were once extracted from oak galls. This is the only practical use that has ever been found for these growths.

bean- or cone-shaped galls. The Hungarian gall wasp (*Cynips quercusalicis*) gives rise to a hard gall on the acorn that tops it like a hat.

In hot areas, a variety of oak gall wasps infects the tree branches. The hard, sharp, sometimes forked outgrowths can be as great as two inches in diameter. The roots of the tree may be infected with andricinid gall wasps; if the tree is young, it will dry up. But woodpeckers will sometimes come to its rescue by opening the galls and eating the larvae.

Finally, the large ball-like galls of the sweetbrier gall wasps have long had a favorable reputation: in Germany they used to be called "apples of sleep," as it was believed that if placed under one's pillow they cured insomnia.

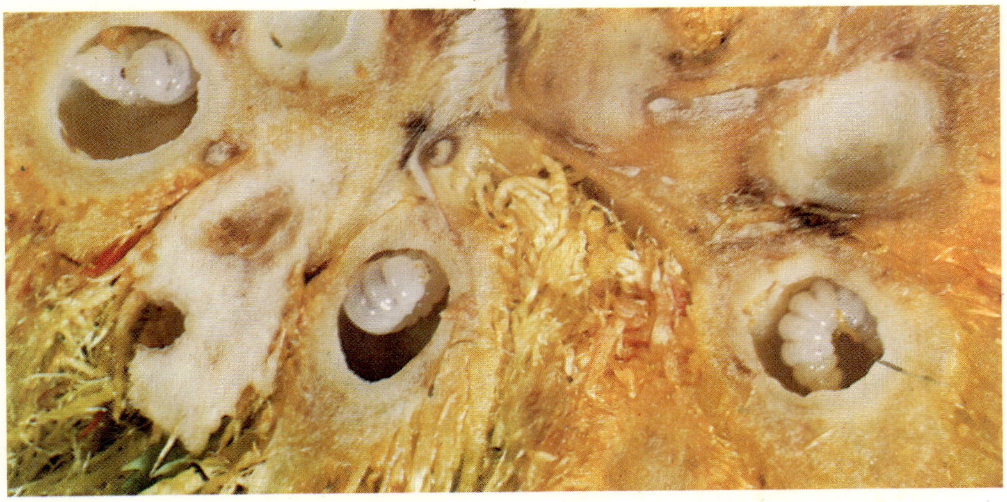

▲ Ichneumon wasp larvae and wild bees sometimes nest together in round galls with larvae of the sweetbrier gall wasps. The nest is concealed by a jumble of long fibrous outgrowths. Here is the old story of the robber robbed: the later arrivals often cramp the development of the first parasites on the scene.

◀ This cross-section of a gall on a rose bush shows each larva in its chamber.

The Hazel Mouse

The hazel mouse is a tiny, appealing animal with red fur and a magnificent tail like a squirrel's. It measures about three inches, with a tail almost that long. It is clever, funny, pretty, and as amiable as a Walt Disney character. It lives wherever there are hazel-nut trees, blackberry bushes, beeches, oaks, and pines, in southern Europe, central Europe well into the heart of Russia, Asia Minor, southern Sweden, and southern England, where it is called the dormouse.

The hazel mouse sleeps during the day and rises with the moon. In a hop, skip, and a jump it is up in a tree, munching on nuts, nibbling on fruit and berries (and succulent young shoots in the spring), as well as on small insects here and there. Its coffee-bean eyes can be seen gleaming from branch to branch.

Once the sun rises, the saunter is over: it's time to go to bed. Like a bird, the hazel mouse builds a nest of cleverly gathered moss, leaves, and tufts of grass, which it hangs from a tree, sometimes six feet above the ground. The nest is a two-to-three-inch ball (or sometimes as large as five inches across if the nest is a family residence). This is a temporary, summer home. The winter nest is large and well padded, with underground warehouses; it is built in a hole or in a tree hollow on the ground. In mid-October the hazel mouse retires to the nest, curls up with tail tucked over head, and hibernates for six or seven months. If the winter is hard and the frost heavy, it risks never waking up in the spring: its body temperature may lower to just above the freezing point and its margin of resistance is greatly reduced.

The first litter is born during the winter in these underground homes. The second litter is born in the branch nest. At seven weeks the babies—usually three to seven—become independent, and in turn begin to build their own nests.

In summer, the hazel mouse (*Muscardinus avellanarius*) sleeps in a ball-shaped nest of grass, moss, and leaves, which it hangs in the trees, three to six feet above the ground. In winter, it builds a spacious and comfortable home in a hole, equipped with storehouses. One litter of young is born in the trees, the next in the ground nest.

Ladybugs and May Beetles

Ladybug beetles (*Coccinella septempunctata*) are sensitive to the cold, and they seek warmth in the winter. Often they wedge themselves in groups between the trunk and the bark of a tree; for the top photograph, the bark was lifted from a twig. Ladybugs hunt green flies and can kill up to fifty a day and even more larvae. The ladybug beetle family includes over 3,400 different species.

As good luck charms, ladybugs are always welcome. But few people realize that the ladybug is useful as well as charming: a ladybug can eat fifty green flies, or aphids, a day, and even more larvae. These spotted beetles are so useful that they are deliberately raised for the purpose of saving especially threatened crops. Furthermore, they reproduce very rapidly. One female lays four hundred to seven hundred eggs on the leaves. After three weeks undergoing all the stages of a complete metamorphosis—egg, larva, chrysalis, adult—these eggs yield hundreds of ladybugs.

The ladybug has only one trick for protection: it can simulate death perfectly. When grasped tightly in the fingers, it becomes immobile: its six legs become rigid, and a yellowish liquid is secreted from its joints. As soon as it is freed, it comes back to life. Come winter, ladybugs seek refuge in houses, or gather in some other warm shelter—a hedge, a thick tuft of grass, the hollow under the bark of a tree—and wait together for the spring.

The May beetle, one of the cockchafers, is as well known to man as the ladybug, although for less happy reasons. Its larvae, which look like white worms, silently destroy everything they find underground in vegetable gardens and nurseries. For its three weeks of life, the adult May beetle is like a little devil with horns that spares no leaf. In what gardeners call a "May beetle year," even keen pursuit by their natural enemies (bats, birds, badgers, hedgehogs, and moles) cannot effectively limit the devastation May beetles cause. These May beetle years recur every four years although, contrary to popular superstition, they bear no relation to leap years. This pattern simply corresponds to the different stages of the insect's development. In the spring, the female lays

▲
Although it looks like a devil with horns, the common May beetle (*Melolontha vulgaris*) actually bears lamella-tipped antennae—seven for males, six shorter ones for females—which have olfactory pits that serve to orient the insect.

sixty to eighty eggs about ten inches deep in the fresh ground. Four to six weeks later the larvae emerge. For three years, while growing and successively shedding their skin until they reach a length of about two and a quarter inches, the white "worms" do nothing but feed intensively. During the summer of the fourth year, they dig a hole three feet deep in which they form a cocoon. At the end of autumn, the May beetle grub emerges from the cocoon but still remains underground. It will appear above ground the next spring, four years after the laying—and will die three weeks later, after having done extensive damage.

The clumsy and heavy May beetle flies only at dusk, when it meets fewer of its natural enemies. To take off, it lifts its wing sheaths and fills up its respiratory tract with air, making its characteristic buzzing sound. ▶

Minks and Polecats

The mink is the symbol of luxury; the polecat, that of stinginess. The mink is the dream of elegant women, while the polecat would make them flee. Yet the two animals are members of the same family, and look so much alike that in the wild they can hardly be told apart. Both are weasels, which, along with the felines, are among the oldest of the carnivores. Minks are native to North America and Europe; the polecat is found only in Europe and Asia. Both havé now been introduced into Australia to control the rabbit population. Both hunt by night and are inveterate carnivores, with ferocious teeth and an appetite to match: they feed on small animals—rats, mice, rabbits, frogs, birds, snakes, and lizards. Slender, lithe, rarely frightened, all their senses alert, they slip through the thickets. They kill not only when hungry, but also store up food in their lair. This food for storage often consists of birds' eggs, which they carry between their chin and chest; but they will also take live animals for this purpose, such as frogs, which they paralyze by biting the nape of the neck at the medulla.

Both species mate in March. Baby minks are born after nine-week gestation period; polecats need only six to eight weeks. After fifteen days of blindness and helplessness, the young develop rapidly. At four weeks they learn to hunt

◀ The cautious polecat (*Mustela putorius*) hunts for food at night, preying on rodents. It is a member of the weasel family. The young learn how to hunt in their first few weeks of life, their cruel teeth already sharp.

collectively. At two months, they leave the family group and live independently.

The mink, which is twelve inches long with six to eight inches of tail, digs its burrow in grassy swamp beds, in reeds, and in dams. Its favorite shelter is a hollow in the roots of an alder, from which it carefully builds several emergency exits into the water. The mink's webbed paws allow it to swim rapidly and for long periods of time. It catches fish, even those as lively as trout or salmon. As minks are hunted mercilessly for their sumptuous fur, they are now very rare in the wild. However, millions of them thrive on breeding farms.

The polecat's fur is darker than the mink's, although its cheeks are lighter in color, making the animal appear to be wearing a mask. Its body is a little larger: sixteen inches long with six inches of tail. It is slower and more cautious. It seems to reflect, avoiding obstacles rather than charging head on. It uses surprise as well as force when attacking. Its character is mean and aggressive, yet it is tolerated by man because of its destruction of rodents. When attacked, its squirts a secretion from glands near the anus which has an unbearable odor. This "ultimate weapon" has saved it from extinction, and polecats are still found all over the small woods cf the temperate parts of Europe.

▲
The more in impetuous European mink (*Mustela lutreola*) charges an obstacle rather than going around it. Hunted everywhere for its precious fur, the limits of its range in the marshes and swamps are constantly shrinking.

Old World Hares and Rabbits

Rabbits and hares look very similar running across a field, though they actually have few characteristics in common. They are neither the same size nor the same weight, nor do they move in the same gaits, nor do they exhibit the same behavior. They do not share the same number of chromosomes, and cross-breeding is impossible.

The brown hare is common all over Europe as far as the Urals, and in North, southern, and western Africa. It has been introduced into Australia, New Zealand, and North America, where it has become acclimatized. It shows a marked prediliction for fields, the edges of forests, and small woods and marshes. Approximately twenty-five inches long, it can weigh up to eleven pounds. The color of its fur varies according to habitat, age, and season, ranging from red to grayish-beige. It always makes its form, or lair, on open, flat ground. When chased, it scampers away at full speed, making sudden turns or going back on its own tracks. Sometimes it will suddenly sit back on its hind legs and survey its surroundings, ears flapping in the wind. Its diet is totally vegetarian: at dusk, it searches out young cereal plants, clover, lettuce, young cabbage, carrots, turnips, and grass; it will invade gardens, which it pillages hungrily when necessary.

The hare's mating season is long, the eight months from January through August. Three or four times a year, after gestation periods of approximately forty-three days, the female gives birth to a litter of one to four, or even six, babies. The young are born with fur, and can see and crawl. This is fortunate for them, for the maternal instinct of the hare is not very strong. The mother spends only the first few days with her young; for the next three

The Old World rabbit (*Oryctolagus cuniculus*) is ▲ never an only child. One litter may have as many as fifteen young, and the female adult gives birth five or six times a year. She is a conscientious mother and protects her young as long as necessary.

◀ Forever digging and eating, rabbits live collectively, sometimes in fairly large colonies. They never stray far from their hole, so that they can always rapidly find cover.

◀ The young European hare (*Lepus europaeus*) is born with two handicaps: its lair, out in the open, is primitive and dangerous, and its mother cares for it only briefly and then abandons it to all sorts of perils.

weeks she returns to them only to nurse them in the morning and at night. For the rest of the time they are left alone, curled up against one another, piteous, helpless, unprotected from the greed of birds of prey, weasels, and foxes. This maternal neglect would be fatal to the survival of the species—each year many young hares die because of it—did nature not compensate: the gestation of a new litter will often begin before the preceding litter is born. Because of this overlapping, the hare population remains numerous, despite carnivorous natural enemies and intensive hunting.

The Old World rabbit is smaller than its cousin the hare; it is only eighteen to twenty inches long. The female rabbit, like the hare, breeds rapidly; she gives birth each year to five or six litters of four to fifteen young, which are born furless and blind. Unlike the hare, the rabbit protects her young as long as necessary. Rabbits are found over a large portion of Europe, and have been introduced into America and Australia. They breed in vast numbers and have an insatiable appetite, ravaging crops and destroying meadows and trees, whose roots they attack from their underground burrows. Australia tried to solve this problem by infecting the rabbits with the myxomatosis virus, which spread rapidly. The operation was a success, and rabbits died by the tens of thousands. But the virus also destroyed domesticated rabbits, descended from the wild rabbits. In Europe, where the virus was also introduced, the results of the epidemic are still evident, and to hunters the remedy seems much worse than the original ill.

The hare's mating season lasts from January through August. The male mates with as many females as possible. In winter, this Don Juan becomes solitary once again. ▶

The Hobby

Of all the birds of prey, the fastest in the air, the most skillful acrobat, and the most intrepid diver is perhaps the hobby, one of the falcons. In design, its streamlined wings resemble those of a supersonic jet. It circles high in the sky, watching, and when it spots a small bird, pounces on it in midair at a speed of sixty-five miles an hour with such violent impact that the prey sometimes falls to the ground, where the hobby finishes it off.

At fourteen inches, the female is often larger than the male; both have a rather large head with a beak whose upper half has a toothlike projection that fits a precisely adapted groove in the lower half. The talons are used as weapons and thus are powerful and highly developed. Before devouring their victim, hobbies

At a speed of sixty-five miles an hour, the hobby (*Falco subbuteo*) dives on its prey. It has no difficulty catching swallows, swifts, and large dragonflies, and often devours them in midflight.

pluck it, except for the wings, which they leave attached to the shoulder girdle.

Since they hunt small birds and dragonflies, hobbies nest exclusively in trees (often in an expropriated crow's nest), in small woods or sparse forests. In October, they leave Europe for equatorial Africa and then return in April.

Four or five eggs are laid annually and brooded for four weeks, by the mother and father alternately. The mother feeds the young for the first fifteen days, sheltering them under her wing. The training of the young begins at five weeks: the parents release baby birds and large insects in front of them, encouraging the young to dive at them and attack.

Old World Robins and Cuckoos

Twittering and hopping about on its long legs, eyes rapidly blinking, the robin is one of the most charming of the singing birds. It is also one of the earliest: in March, it announces its return from warmer climes by singing piping arpeggios in European forests. But however lovely the song may be, it is also a warning: this is how the robin announces the limits of its territory, which other males are not allowed to enter. On the other hand, the

female is perfectly welcome; the male greets her with a little whistle and then patiently waits for her to respond to his advance.

The female builds her nest close to the ground, or even in a thick bush at ground level, or in a well-concealed hollow under an old willow. She carefully constructs a roof of moss and feathers. She lays four to six eggs twice a year, so close together that she is brooding the second clutch while the father is still searching for food for the first. The babies grow up peacefully in the warmth and security of the nest—unless their family life is destroyed by a cuckoo. This relationship between the two species is one of the strangest enigmas of animal life.

The female cuckoo evidently does not like to be hampered by family responsibilities. So, with great proficiency, she practices the art of substitution. In the

Baby robins are born in a well-concealed nest ▲ either very close to or on the ground. They are brought up by both father and mother. At fifteen days, they begin to fly. Two clutches of eggs generally follow directly upon one another; thus one will sometimes see the babies of the first and the eggs of the next in the same nest.

Opposite:
Worms, insects, and small snails are the usual diet of the European robin (*Erithacus rubecula*).

Sixty different species of the cuckoo's host birds have been identified: these are most often robins, but sometimes chats, warblers, green finches, and even tiny kinglets. The female cuckoo always chooses the species she herself was raised by to raise her own young, and lays eggs very similar in appearance to theirs.

Robins will sometimes brood a cuckoo's egg that has been substituted for their own. The young cuckoo (*Cuculus canorus*) will throw its adopted siblings out of the nest and monopolize the parents' attention.

early morning, while the female robin is brooding, the cuckoo is watching carefully. Whenever possible, as soon as she sees the robin parents leave the nest in search of food, she quickly lays an egg on the ground, picks it up in her beak, and rapidly slides it into the robins' nest. She removes one of the robin eggs and throws it violently to the ground. The red-specked white eggs of robin and cuckoo are very similar in appearance. Twelve and a half days later, the baby cuckoo hatches in the robins' nest, bald and spiteful. It crushes the robin eggs one by one with its stubby wings, and kills the live baby birds the same way, and throws them to the ground. The robin parents seem blind to all this, as if mesmerized by the gaping beak of their giant new child, they feed it with surprising readiness. Even if one of their own offspring is pitifully hanging on to a branch near the nest, they no longer seem to recognize it, and ignore it completely. The young cuckoo leaves the nest when it outgrows it, but the robin parents continue to feed it for three weeks.

Laying one egg at a time, the female cuckoo repeats this procedure twenty times over a period of thirty-five to forty-six days. The cuckoo always chooses the species of bird—most often the robin—that raised her to raise her own young, and she always lays eggs identical to those of the adoptive parents. Apparently she does not always manage to find homes for all her eggs, but a sufficient number manage to survive to insure the continuance of the species, and to destroy countless nestfuls of robins.

The hunger of the cuckoo is still voracious in adulthood. The cuckoo devours great quantities of forest parasites and pests, particularly processionary caterpillars, which are feared by other animals because of their stinging hairs. Thus there is a balance between the harm and the good the cuckoo does.

8. Fields and Meadows

European Fence and Wall Lizards

Dependent on heat and light, these lizards are obsessed by one thing in life: finding the sun. They live in scattered meadow glades, dry fields, and vineyards, on walls and on rocks—anywhere they can find the sun's rays.

The activity of these sun-worshipers begins at dawn and ends only at dusk. They spend the whole day langorously watching for insects, which they overwhelm in a surprisingly rapid dash. They use the tongue to test the environment. Extended out of the front of the mouth and retracted, it picks up minute particles that make up odors. Sense organs located in the palate then identify these, thus complementing the olfactory information transmitted through the nose. When the lizard spots prey, its hurls itself on its victim and crushes it with its tiny teeth. Like birds, the lizard has a nictitating membrane, a sort of third eyelid. The lizard is covered with small body scales, and larger oval scales on the tail. Each leg has five clawed toes; those on the hind legs are of varying lengths. These enable the lizard to climb any rock or bush skillfully.

These lizards feed on spiders, insects, caterpillars, and small worms, which it can catch on the driest ground. The prey is snatched up by the jaws, crushed superficially, and then gulped down. The lizard gets its water by drinking drops of dew. When threatened, it instantly vanishes under a rock, in a wall crevice, or into a pile of brush. The lizard's tail is extremely fragile, and fractures easily; the ability to break off its tail when seized is an advantage to a lizard in danger, since the wriggling movements of the detached tail may distract a pursuer and give the lizard time to escape. The lost tail is soon regenerated.

The fence lizard, whose body temperature depends on external circumstances, follows the cycle of the seasons. In cold weather it hibernates in groups under a pile of dead leaves or moss, or in a small hole. But if the winter is very harsh, or if the lizard doesn't bury itself deeply enough, it may freeze and die.

◀ Green lacerta lizards (*Lacerta viridis*) spend hours immobile in the sun. Sixteen inches long, they are one of the largest European species.

The green lacerta lizard, a good climber, catches the heat and light of the sun from tree branches. When it must flee, it jumps to the ground, often landing several yards away from the tree. ▲

◀ In spring and summer the European fence lizard (*Lacerta agilis*) displays its brilliant green sides. Its dress of brown or gray scales is highlighted by a dark horizontal stripe.

▲ To a lizard, losing a tail is not a traumatic experience. This European fence lizard has lost its tail to a pursuer, but it will grow back.

The fence lizard mates immediately after hibernation. The mating season is short, and once mating has occurred male and female separate, each anxious to establish its own territory. One related species is ovoviviparous; the others lay five to fourteen parchment-like eggs in moss or powdery soil. The female wall lizard cleverly chooses to lay near a black ant's nest, which will provide generous quantities of food for the young. Depending upon the species, lizards hatch after an incubation period of four to twelve weeks; they are barely an inch and a half long at birth and must immediately provide for their own needs. Their enemies are many—including their own parents, who may eat them.

The European fence lizard is under legal protection. The wall lizard, which has lived over the ages in the neighborhood of man, in vineyards and ruins, can be as long as sixteen inches. It always seeks the sunny side of hills, rocks, or walls. An impulsive hunter, it is one of the handsomest and most fascinating animals of the reptile world.

◀ This young lizard, just hatched, faces many dangers—including that of being eaten by its own parents. The heat produced by the fermentation of the soil assures incubation of the eggs.

▲With its dazzling colors, the green lacerta lizard is considered one of the jewels of the reptile world. Like all its kind, it sheds its skin several times each summer.

The European fence lizard undergoes similar moltings. While snakes shed their skin all at once, leaving behind the characteristic one-piece coat, lizards lose theirs in pieces by rubbing up against rough surfaces. ▶

The Hamster

Captivity sometimes subdues animals. The pet hamster, a domesticated form of the golden hamster of Syria, is the sweetest and most entertaining of companions. In freedom it is surly and aggressive, as is its somewhat larger relative, the common hamster, which is found throughout the meadows and steppes of Eurasia, showing a marked preference for wheatfields.

The common hamster is a burrowing rodent. It is approximately twelve inches long, with an inch and a quarter of tail. Its soft and thick body fur is tricolored, black, orange, and brownish yellow, with a white throat and white legs; its eyes shine like small black pearls. Both male and female live alone; each builds its burrow three to six feet underground in loam or loess and lines it with hay and grass. The layout of the burrow is complex: there are storage areas, an adjacent hole for excrement, access corridors, and a vertical exit for strategic retreats in case of emergency. During the summer, the provident hamster builds up its reserves of food: wheat, peas, beans. It carries them in its cheeks in loads of three and a half ounces, and when its mouth is thus stuffed its head is three times the size of its body. By autumn, it has accumulated about a hundred pounds.

When the cold months come, the hamster hibernates. Its body temperature drops from 93° F. to 39° F. Every three weeks, the male wakes up and eats copiously. The female, because she was caring for her young during the summer and could not store up as much food, goes into a deeper hibernation and sleeps without interruption.

Mating occurs two or three times a year. Immediately afterward, the female bites the male to chase him away. Twenty days later, six to fifteen babies are born, hardly larger than worms; they can see at eight or nine days. In preparation for their first ventures outside the nest, the mother digs several more vertical corridors to the burrow, thus making it possible for the young quickly to escape buzzards, falcons, foxes, polecats, and weasels. This is her final maternal act. At fifteen days, young hamsters leave the burrow and learn to struggle for life. This may account for their irritability in the wild and their amiable nature in captivity, where they are safe from danger.

This puffy-cheeked common hamster (*Cricetus cricetus*) has started gathering its stores for the winter. It stuffs up to three and a half ounces of wheat or dried vegetables into its mouth on each trip to its burrow.

Opposite:
Brave but not reckless, the hamster always checks its surroundings before venturing out, turning its black pearl-like eyes in all directions.

Paper Wasps and Hornets

The angry buzzing of a wasp or hornet has disrupted many a pleasant picnic—these insects, especially the females, sting viciously.

Paper wasps are as elegant as they are vengeful, with their yellow and black rings, their delicate structure, and their gossamer wings; together with hornets, which can be as long as an inch and a quarter, they belong to the family Vespidae. This family includes three hundred different species, of which ninety are European and at least as many are found in the New World. They feed on nectar, but will also hunt spiders and other insects, including

◀ This remarkable document illustrates the different stages in the development of the common hornet (*Vespa vulgaris*) in the chrysalis. The larvae have developed from eggs in the hexagonal cells. The workers feed them a gruel of premasticated insects. When the larvae are complete, they seal their cells with a wax cover and weave a glassy cocoon in which they become chrysalises. The hornet develops in the cocoon, which it leaves fifteen days later as an adult.

▲ Hanging from a branch, or tucked under the edge of a roof or the overhang of a rock, the nest of the French paper wasp (*Polistes gallicus*) is protected by a sheath. This species is harmless.

The upside-down larvae and chrysalises are anchored by their abdomens to the roof of their ▼ cells.

their own kind, which they chew into a premasticated gruel for their larvae. Their powerful mandibles seize the prey, often in midflight. The inedible parts, such as the wings, which are coated with chitin, and the legs are bitten off, and the rest of the body ripped apart. When fed, the larvae offer a drop of fluid to the workers in exchange; the workers are so eager for this liquor that they encourage the larvae to secrete it at other times as well as at feedings. Though this draining may weaken the larvae, the adults seem to thrive on the fluid.

◀ Raiding and stinging are constant activities of the German hornet (*Vespa germanica*). It is three-quarters of an inch long, and its shieldlike head has one to three white spots and its eyes are surrounded by yellow rings.

A social wasp colony, whose numbers vary from a hundred to fifty thousand, lives only one year. In the autumn, female wasps leave with the males on their nuptial flight. After the flight the males die, and the fertilized females go into hibernation. In the spring, each wakes up a queen that will start her own colony. The first larvae hatch in the heat of the sun and immediately begin building the colony's cells out of wood fibers to which they add their own saliva. The queen continues to lay in these cells, and she cares for the second generation herself, too. After this generation she will concentrate solely on laying, since there are now enough workers to care for the larvae and continue the construction and expansion of the nest. The larvae are placed in horizontal furrows protected by several layers, connected and supported by tiny shafts and divided into hexagonal cells with the opening at the bottom. The larvae remain head down, attached to the cells by the abdomen. At the end of the larva stage, they seal up their cells and spin glassy cocoons in which they are transformed into chrysalises. At the end of fifteen days, young wasps emerge, generally workers. They are first employed on the inside of the hive; when they are more experienced, they are responsible for gathering honey outside. In autumn, the queen and her workers die. The cycle begins again as the reproducing wasps make their nuptial flight and found colonies the following year. Sometimes, however, in hot summers when the plants are rich in nourishment, the queen survives and hibernates through the following winter, along with some remaining fertilized eggs. These will develop in the old nest and begin a new colony.

Some wasp nests are built in the shelter of a garage or any overhanging projection; others, like the ball-shaped nests of some hornets, are suspended from trees. The nest is always built of wood fibers and saliva, and looks like molded cardboard.

Some hornets and yellow jackets establish colonies of a thousand to four thousand members, building their nests underground in rotten wood. They also inhabit tree hollows or abandoned nests, to which they leave a passage no larger than their bodies. Their sting is painful. Few wasps are as dangerous: a man attacked by several of these insects can die.

▲ Destroying a hornet nest invites immediate retaliation. The nest is frequently found in a natural hollow or a mousehole.

▲ The Saxon hornet (*Vespa saxonica*) protects its colony by building its nest under a roof or ledge and sealing it off with three protective sheaths.

▲ The Saxon hornet, seventeen millimeters long, has a shieldlike head and black horizontal spots. This nest has been opened to reveal a larva.

◀ The female European hornet (*Vespa crabo*), which has been introduced into North America, is the largest of the Vespidae. She is almost an inch long, and attacks large insects as well armed as herself. In some circumstances her sting can be fatal to man.

The wasp, although fond of ripe fruit, hunts insects as well, which it prepares into a pulp to nourish its larvae. ▼

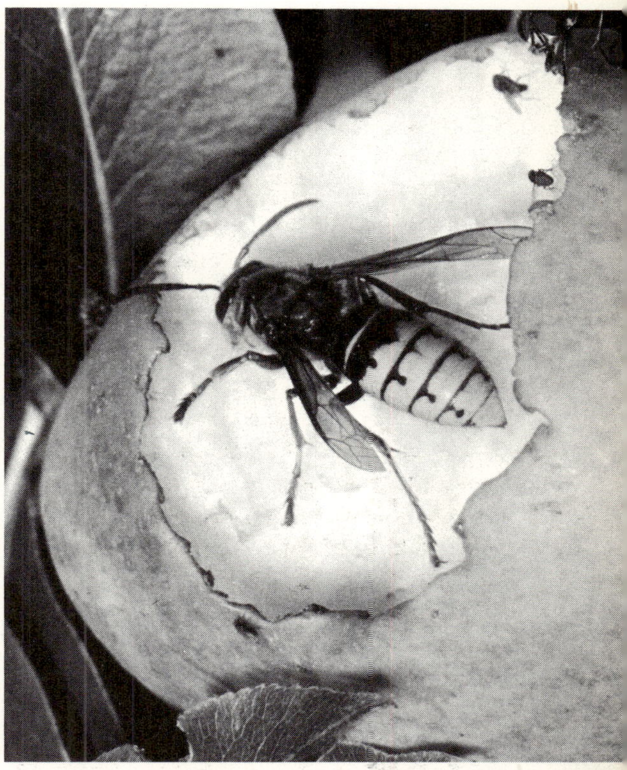

Its abdomen striped yellow and brown, *Vespa crabo* is recognized by its disproportionately large head and black thorax. Colonies of wasps live only one year. ▲

Butterflies

▲
These kaleidoscopic painted ladies (*Vanessa cardui*) fly over the Alps each year, leaving North Africa or southern Europe for temperate climates. The second generation returns south in the autumn.

Above:
A European blue butterfly in silhouette.

They look to us like winged jewels or dancing rainbows, whose only role is to dazzle our eyes. Nature has scattered butterflies over all the continents except Antarctica. Most feed on flower nectar and pollen, although others absorb nothing in their ephemeral adult lifetimes. The brief period—a few days, sometimes only several hours—between emergence from the cocoon and death is used for one activity alone: mating.

While its night relatives, the moths, are thick-bodied, diurnal butterflies have a thin streamlined body and two knob-tipped antennae. Like moths, they seem to be aware of the dangers of displaying their beauty, and when at rest they fold their speckled wings back vertically, thus exposing only the dull underside. Some hibernate, and can freeze and become as hard as ice without dying; others die in the autumn. Still others (although this is true of more moths than butterflies) are migratory; Old World species arrive each

In winter, the dazzling peacock *(Inachis Io)*, one ▲ of the most beautiful butterflies found in the Old World, seeks shelter in buildings and sheds.

◀ Seen under a microscope, the wing of a butterfly, like the moth's wing, resembles a mosaic of fine scales arranged like the tiles of a roof.

The cress butterfly (*Anthocaris cardamines*) has ▶ triangular wings the color of the rising sun. The undersides are veined with green.

▲

Frost does not mean death for the lemon butterfly (*Gonepteryx rhamni*). This butterfly hibernates on the ground, between dry leaves, and can become as brittle as ice and still survive.

summer from North Africa and the Mediterranean countries, while New World varieties fly north from Central and South America.

But this lovely thing is a caterpillar before it is a butterfly. The female instinctively lays her eggs on the plant that is the most nourishing for her larvae, on the backs of leaves, in the clefts of bark, in coils around branches, sometimes covering them with a protective net of bristles and splinters. The hairless or nearly hairless caterpillars are clumsy, their tiny eyes barely allowing them to discriminate between darkness and light. They limp

◀The emperor (*Argynnis paphia*) is often found in forest glades. This species mates in the air, the male carrying the female away on his back.

▲
The red admiral (*Vanessa atalanta*) has a difficult migration over the Alps. The return voyage south is often hazardous.

along, as if their three pairs of true legs attached to the thorax and the additional five pairs of pseudolegs under the abdomen were not sufficient to bear their long body. They move the legs on the thorax first, steadying themselves on the abdominal legs whose hook-covered soles allow them to grip the leaf or branch surface solidly, and thus move the entire body along, pushing off from the last pair of back legs. Generally a thread of silk is played out as the caterpillar moves along; this acts as a safety line.

Caterpillars rarely, sometimes never, leave their feeding plant. Their appetite is

▲
Cabbage butterflies (*Pieris rapae*) are so named because they are born in cabbage plants. The female lays her eggs on the backs of the leaves.

▲
Fifteen days after laying, tiny two-millimeter-long caterpillars hatch. As soon as they are strong enough they separate and go their own ways.

▲
In May the adult cabbage butterfly leaves its cocoon, which remains stuck flat to the ground. A new cycle now begins.

enormous. Their initial weight very rapidly multiplies by a thousand, which means that they must shed their skin several times before they attain their final length. Shortly before the resting stage during which they wrap themselves up in their cocoons, caterpillars store the additional food necessary for their metamorphosis. Then, in a sheltered spot, they weave their cocoons. These often look like tiny sarcophaguses—but it is a gloriously alive creature that will emerge from the darkness.

With their border of black and turquoise and their striking design, the wings of this swallowtail butterfly (*Papilio machaon*) seem to have been painted by the irrepressible brush of an artist. ▼

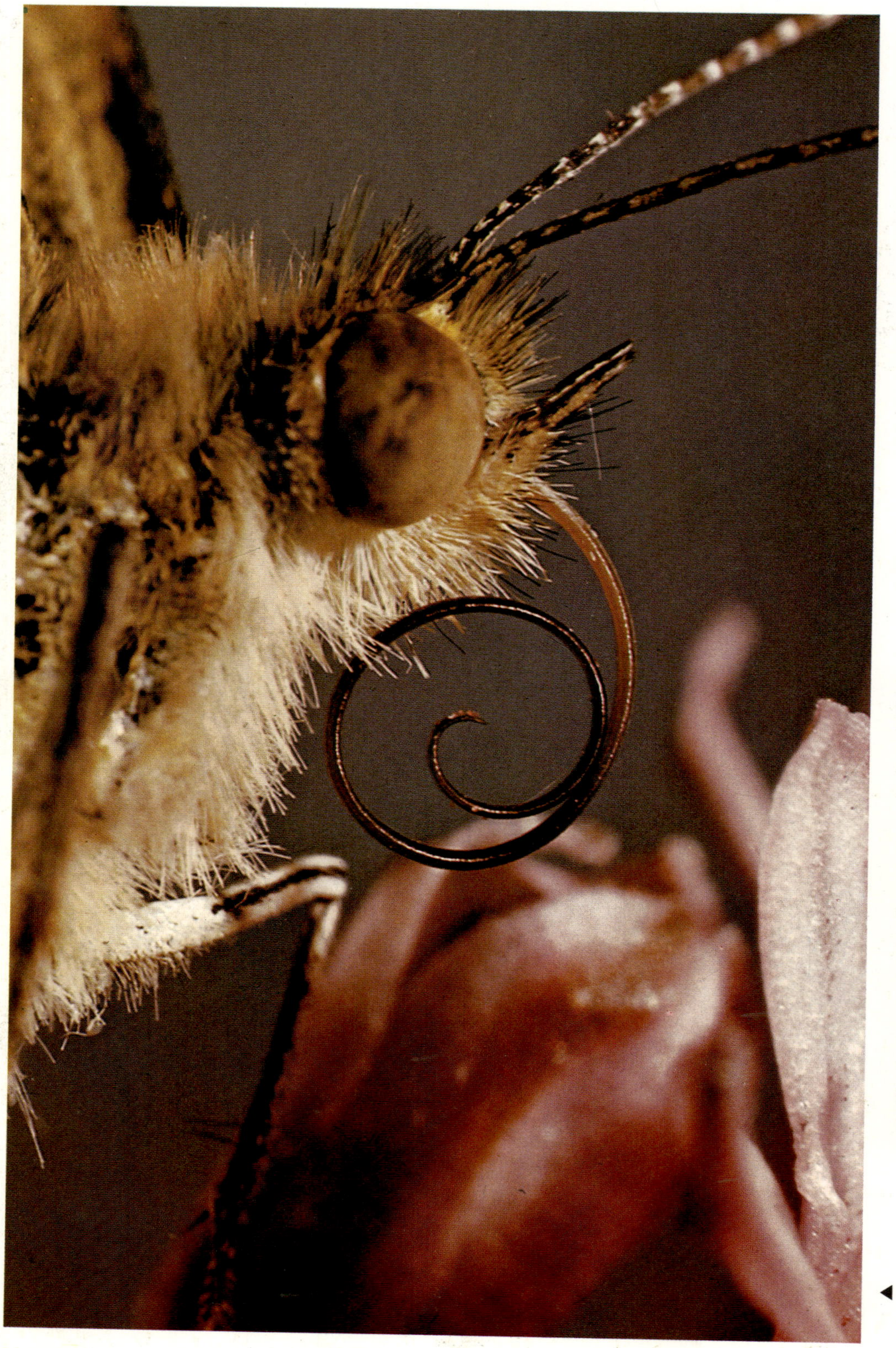

◀ Describing a mathematical spiral, the proboscis of this European blue (Lycaenidae) is rolled up under the head. ▶

The Buzzard

▲ Perched on a stake or bush, the Old World buzzard (*Buteo buteo*) watches for lizards, snails, and beetles; its main diet, however, is rodents, which it hunts from the air.

Without once beating its wings, without a single sudden motion, the buzzard circles beautifully for hours like a glider set forever on a 360° course. The best known and most common of the European diurnal birds of prey, it defines the limits of its territory with these magical circles. It chooses areas of open countryside, wooded valleys, and scattered small copses as its hunting grounds. Sometimes flying high in the air, sometimes skimming close to the ground, it watches closely for its prey. Its sharp eyes rapidly spot the mice, moles, hamsters, or weasels that make up its usual diet. It dives down from its circling, wings flatted out against its body; it suddenly brakes, seizes the prey, and shatters its skull or spine with one blow of its hooked beak. The curved upper half of the beak overlaps the lower half, thus making this weapon even more lethal. Small victims are gulped down whole and even their bones are digested, while the fur, feathers, claws, and teeth are regurgitated in a wad. Sometimes, in order to vary its activities and menu, the buzzard changes its hunting techniques. It will perch on a gate post, or an isolated tree or rock, and practice its skill at catching lizards, grasshoppers, snails, or beetles.

During the mating season, the male and female fly in pairs: they let themselves sink vertically through the air together, coming toward each other; they avoid collision by gracefully swerving; then, wings outspread, they spiral back up again. The female lays three or four eggs in a large (thirty-inch), crude nest on the edge of a forest in a solid tree fork, on a bed of briars, moss, fur, and feathers. For thirty days father and mother take turns brooding the eggs. When the gray- and white-flecked babies are born, the male searches for food and the female distributes it. At six weeks the novices begin to learn to fly and circle like the adults. When mature, they measure twenty-two inches and their wingspan reaches four feet. The same nest, enlarged and improved, is used for several generations.

Some buzzards are migratory. In autumn they fly south and return in March. Most, however, prefer remaining in their native land. If the winter is very harsh, and especially if it is snowy, many die. The strongest are forced by necessity to attack game such as pheasants, partridge, or rabbits.

Buzzards are under legal protection, as they destroy rodent pests. They are easy to recognize by their flight patterns, and their cry is even more distinctive: they meow like cats.

◀ The gray- and white-flecked babies leave the nest at seven weeks. Before this, the parents have taught them to fly and circle.

▲Crude but capacious and sturdy, the nest, thirty inches in diameter, is always found high in a solid tree on the edge of a wood. The female furnishes it with briars, moss, fur, and feathers.

◀A proud bird of prey, the buzzard measures over twenty-two inches and has a wingspan of four feet. Its curved, pointed beak, whose upper half overlaps the lower, is well adapted for crushing animals.

The Cross Spider

Spiders are found all over the world, in both hemispheres, in cities, jungles, deserts, and even in the highest snowy peaks of the Himalayas. They are beneficial to man, particularly to gardeners, since they prey on pest insects. Their fantastical bodies and their dewdrop-jeweled webs are the stuff of fairy tales.

The cross spider, a marvelously gifted spinner, is one of the most outstanding of the orb weavers, but its talents could not

◀ A master weaver, the cross spider (*Araneus diadematus*) builds a circular web hanging in tree branches or bushes. A warning strand connecting the center of the spider's "home base" at the edge of the web warns it of the presence of a captured insect.

Pearled with dewdrops, the web sparkles in the sun. This fairy veil is made up of forty thousand strands of thread, each no thicker than one- or two-thousandths of a millimeter. ▶

be truly appreciated until the discovery of the electron microscope. Today, this instrument has shown us that the cross spider's web is made up of forty thousand strands of thread, each one- to two-thousandths of a millimeter wide; the finest thread produced by man—fiberglass or tungsten filament—has a diameter of five-thousandths of a millimeter. The spider's abdomen is like a small factory: three pairs of spinnerets on its tip bear a "loom" dotted with forty thousand incredibly fine nozzles, each of which emit one of the sticky threads; these are wound on bobbin-like structures. From this web no insect prisoner escapes.

But the spider has not yet completed all the architectural refinements of its web: it secretes other, different-sized threads, which are not sticky. Some of these are used as bannisters, others as bridges, and yet others to tie up a victim—paralyzed by the spider's bite—and to hang it in the shade in readiness for a future meal. The spider even weaves an alarm thread that connects the center of the web to the spider's retreat at the edge of the web and signals a catch. Although the spider has poor eyesight, it has a highly developed sense of touch; and its field of perception is enhanced by the circular web. A web 140,000 millimeters square, for example, provides a sensory amplification comparable to a man's view of an object through a lens with a magnification of 154.

The spider does not actually swallow its prey, but injects its saliva into the captive insect and aspirates the liquid that results from the digestion of the victim's organs thus begun.

At the end of autumn, the cross spider lays a large quantity of yellowish eggs in her lair at the edge of the web and, naturally, weaves another web to cover them in a sort of cocoon; this is her last masterpiece, for she does not survive the winter. The next spring a new generation of spiders emerges from the eggs. Before reaching adult size, the only metamorphosis they undergo is a series of skin sheddings.

The spider suitor needs quick reflexes, for if he is not careful his partner will devour him as soon as mating has taken place. ▼

Ichneumon Wasps

▲ An ichneumon wasp has laid its egg in this privet hawk moth chrysalis. Now, instead of a moth, a wasp will emerge.

In the insect world, which certainly has its share of odd phenomena, the ichneumon wasp is perhaps the strangest of all. There are about ten thousand known species, almost all of them found in the temperate zones of the Northern Hemisphere.

At first sight, the description of the ichneumon wasp seems perfectly ordinary. Size: 1/5 to 1 1/5 of an inch; color: neutral; antennae: long, thin, mobile; wings: membranous; family: Hymenoptera (ants, bees, and wasps). In short, a seemingly uninteresting wasp. But its peculiarity is not in its appearance, but in its behavior. The female ichneumon wasp needs a live incubator in which to lay her eggs, and for this purpose she chooses the larva of another insect. Ichneumon wasps with short ovipositors choose the larvae most convenient for them, those that live in the open air. They inject their ovipositor into the coat of fat around the larva. Neither the best hidden caterpillar nor the smallest louse escapes it. No suitable host is spared: even a chrysalis will do. Ichneumon wasps with long ovipositors seek out larvae buried in the trunks of trees; they

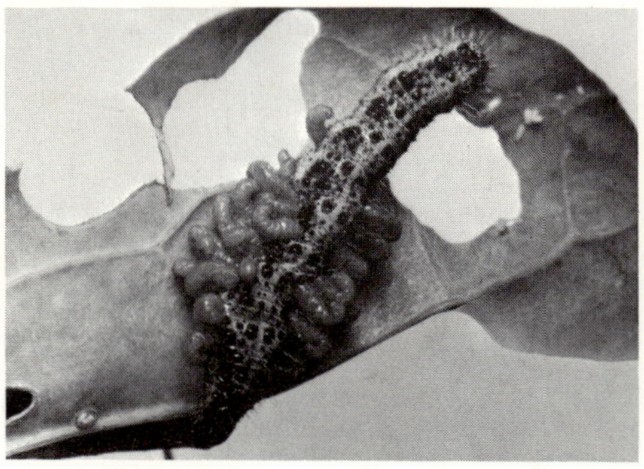

▲ A single cabbage butterfly caterpillar has been used as a live incubator and food supply for the larvae of more than a hundred braconid wasps (*Apanteles glomeratus*), relatives of the ichneumon wasps. These larvae will become chrysalises.

drill into the wood until they reach the grubs and inject their eggs directly into the larva's body. Warm and protected, surrounded by its living nourishment, the ichneumon larva has only to let itself live. It prudently consumes first the host's superficial fatty layer, being careful not to attack the vital organs. Thus, for a while the host continues to develop, though racked by ever mounting hunger. It soon becomes thinner and thinner and increasingly feeble. It actually wastes away, its agony prolonged, until sucked dry by the imbibing parasites within. As the ichneumon larva grows, its sucking organs become modified, and the parasites now emerge from the victim's skin and pupate, using the host's empty skin as a base, and become chrysalises. In some species, the larvae will remain in the shelter of the host's body and pupate there, emerging only as adults.

Sometimes the development of an ichneumon larva is disturbed: the parasite may itself be host to yet another species of ichneumon. The latest arrival first exhausts the earlier guest and then attacks the original victim—providing a tasty morsel for a woodpecker who may come along and swallow the whole thing.

◀ The thin, delicate abdomen of the ichneumon wasp ends in a giant pointed ovipositor. It sinks this into the fatty mass of the caterpillar and lays its eggs. The host has not felt the slightest sting—but it will soon waste away, its vital organs gradually sucked out by the parasite.

An infallible instinct causes each variety of ichneumon wasp to pursue a specific kind of larva: here the host is a butterfly chrysalis. According to some researchers, a chemical substance triggers the hatching in the single favorable host. ▶

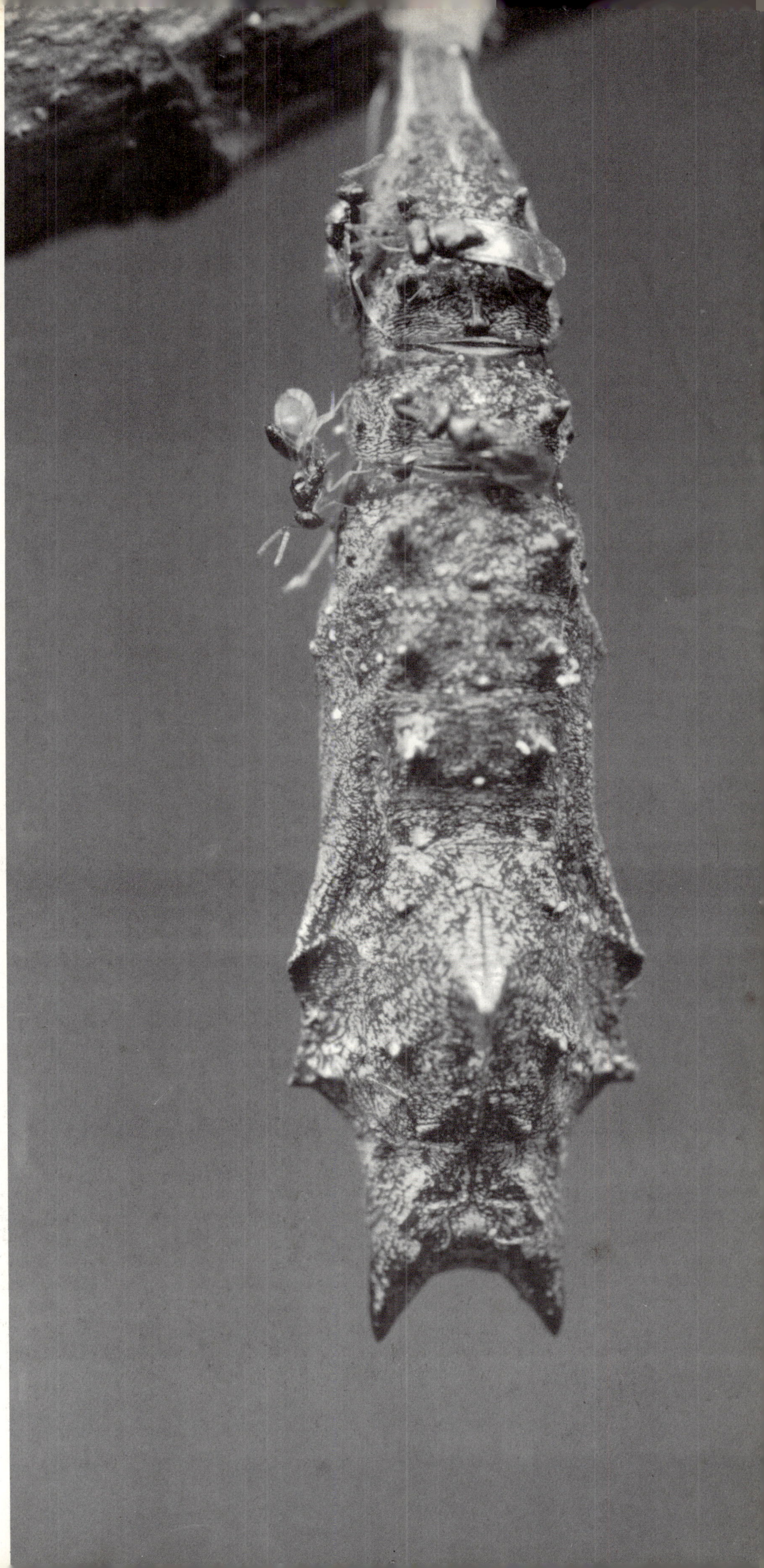

Grasshoppers and Crickets

No meadow or wood would be the same without the incessant chirping of these small, armored creatures. The leaping, peaceful grasshoppers play a loud part in the grassy symphony.

The katydid, a long-horned grasshopper, sometimes mottled with red, measures an inch and a half and can leap two hundred times its own length; propelled by its long back legs, it can jump twenty feet in one bound. When young, it feeds on plants; then it moves on to a meat diet, attacking insects. At the end of the summer, the female drills into the loose soil with her ovipositor and deposits her eggs. The larvae hatch the following spring: they differ from their final adult appearance only in their lack of wings.

The chirping of long-horned grasshoppers is not a song at all, but rather an instrumental piece. It is played only by the males, who attract the females this way. When the grasshopper chirps, its wing sheaths can be seen moving rapidly; fine lamellae located under the left wing sheath rub up against a rib in the right wing sheath, like a violin bow on a string. The sound thus produced

◀ The final act of a grasshopper's life: at the end of summer, the female drills the soil with her ovipositor and buries her eggs deep in the ground, surrounding them with moss, which soon hardens. The next generation is now assured, and both parents die.

Females have auditory organs at the base of their ▶ front legs that enable them to hear the males' mating call—shown here is a European field cricket (*Liogryllus campestris*). This musician rubs the lamellae of one wing like a bow against a rib on the other, which acts as a string.

▲ Front legs shaped like shovels and a heavily armored thorax aid the European mole cricket (*Gryllotalpa vulgaris*) in building long underground passages.

A katydid (*Tettigonia viridissima*). ▼

causes the vibration of a taut membrane over an airfilled cavity at the base of the right wing sheath. This is called the drum, and it acts as a resonating chamber.

The dark, secretive, elusive cricket rarely shows itself—but it is often heard. Its music carries great distances on summer nights, when the male is sounding forth his mating call. The female shows a distinct maternal instinct, which is rare among insects. She not only carefully prepares the holes in which she will deposit her eggs, whose development takes three weeks, but she also waits for her young to mature and disperse—about a month—before resuming her usual occupations. Field crickets live in small holes which they dig themselves on the edges of sunny meadows; at the slightest sound they rapidly retreat into these holes.

The mole cricket can be almost five inches long; it has large front legs which it uses as a shovel, and a heavily armored thorax. It digs long underground corridors, and it feeds on roots. But it also eats grubs, beetle larvae, and granary caterpillars, and is thus a friend to man.

Locusts

For centuries, locusts and disaster have been synonymous; a plague of migratory locusts is indeed a calamity. In the Old Testament the prophet Joel described "a day of darkness and of gloominess, a day of clouds and of thick darkness, as the morning spread upon the mountains. . . . The appearance of them is as the appearance of horses; and as horsemen, so shall they run. . . . Like the noise of chariots on the tops of mountains shall they leap, like the noise of a flame of fire that devoureth the stubble. . . . The earth shall quake before them; the heavens shall tremble, the sun and the moon shall be dark, and the stars shall withdraw their shining." In his *Wonders of Animal Life*, R.-H. Francé relates, less poetically but in identical imagery, his experience of a locust invasion of a Hungarian town: "Preceded by a metallic crackling, like an army of mounted horsemen, the pale yellow cloud came into the center of the street. They were descending in blasts of wind. Before one could even comprehend what was happening, one's head, shoulders, hands, and face were covered by the insects; they were climbing all around, excited, falling all over. The street was one mass of buzzing, leaping insects, and one couldn't take a step. They crushed each other so that a slippery smelly gruel of jumping bodies soon covered the ground. They crawled all over people's clothes and bodies, pretending to bite. The whole thing took only five minutes; then, eerily, the entire swarm left, suddenly and violently, as quickly as it had come, with a bizarre rustling sound like thousands of moving fans." According to Francé, a total of fifty billion locusts arrived in successive waves all through the afternoon.

It is no coincidence that the first recorded locust plague fell on Egypt, for Old World migratory locusts come from the dry sandy soil of the tropical and subtropical plains. Multitudes of females deposit their fifty to ninety daily eggs in the ground, wrapping them in groups of approximately forty in a mossy and sticky envelope. At nine months the young locusts are ready to move. They cover the ground of entire regions: their only goal is food, their only effect, devastation. Still

◀The armored head of the migratory locust (*Locusta migratoria*), equipped with strong mandibles, reveals the insect's fierceness and voracity. For once, appearances are not deceptive.

wingless, they gather in bands and then leave in hordes on their campaigns. Traveling in leaps and bounds, at first they cover about half a mile a day, and soon about six miles. When they encounter an obstacle, the entire army halts in one giant teeming mass, and forms ramps and bridges with their own heaped bodies. When men set fires to try to turn them away, these are extinguished by the mass of charred locust bodies, then the survivors pass over the dead and continue on. After their fifth skin shedding, locusts grow wings: now truly comes the apocalypse.

Locusts can survive eighteen hours in water, and are unaffected by long periods without food. If the head is severed from the body, for twenty hours the body will still be hopping about and trying to mate. Even chemical products sprayed from planes have no effect on them.

Wherever the flying army lands, nothing remains but devastation and ruin. The flurry of thousands of legs, the crackling of wings, the grinding of crushing jaws that pulverize everything in their path, can be heard from great distances. Even chemistry has been unable to conquer this apocalyptic cloud.

9. Rocks and Caves

Swifts and Swallows

Urban dwellers who happily hail the return of spring when they see what they believe is the first flock of swallows are generally mistaken—these swallows are usually swifts. Swifts are larger and wilder birds, in no way related to swallows.

Swifts are interesting birds in their own right. The swift does everything in haste, including flying—it can achieve a speed of 130 miles an hour. It is built like a streamlined plane, with eight-inch wings as long as its body, and a strong, muscular thorax. Its return voyage from the warm countries in the spring is quite a race.

The swift shows the same haste in seeing to the comfort of its young: any nest, vacant or not, will do. The parents cover the eggs or the baby birds with whatever materials they can find on the spot, coat the entire nestful with a layer of viscous saliva, and for twenty days the male and female take turns sitting on the two or three eggs. After hatching, the babies are fed only insects for the first six weeks. Scarcity or temporary lack of food does not necessarily mean death; like hibernating animals, swifts can survive for some time with a lowered metabolism and body temperature.

Swallows, which are small, sweet, and faithful, return to the same nest each year, a construction of mud and clay hanging from a roof or ledge, or even in a chimney. The barn swallow seems to like human company and nests in man's buildings. It generally raises two nestfuls a year, each made up of four or five eggs. As soon as the young can fly, they serve an apprenticeship during the day, returning to the nest at night to sleep with their parents. At the beginning of autumn, swallows gather in large assemblies in the reeds and form communities. Then they fly south in October, and when they return spring has truly come.

▲ Bank swallows (*Riparia riparia*), faithful like all their kind, dig their tube-shaped nests in a steep bank, clay pits, or a cliff. The size of the nest varies from twenty-four to seventy inches.

Above right: The common swift (*Apus apus*), which is brown or black and has long wings, first announces the spring.

This bewildered-looking young swallow, fed nothing but tiny insects, cries for its parents when it is hungry. ▶

The fortunate swifts have a protective mechanism ▶ aginst undernourishment. When food is scarce, they fall into a sort of hibernation that allows them to resist starvation and survive.

Excellent masons, barn swallows (*Hirundo rustica*) ▲ mix a mortar out of mud and straw to build their nest. They raise two or three nestfuls a year. Their frail babies suffer from a lack of food when the summer is particularly damp and cold; some don't live to see the autumn, and others are too weak to survive the long migration south.

Kestrels and Peregrines

The kestrel, a very likable bird, lives halfway between man and the skies. The sky is its unlimited horizon, but today this forest dweller is gradually approaching man, living in towers and ruins as well as copses and quarries. The location of the nest seems not to matter as long as it is high.

As the kestrel, an Old World species, soars over fields in its calm gliding, it always keeps its head tilted watchfully toward the ground. It can spot the smallest mouse from quite high up, and swiftly dives for it. Field mice make up 85 percent of its diet; the rest are mainly lizards, frogs, large beetles, and grasshoppers.

The kestrel's evident enjoyment of space and freedom does not necessarily mean a taste for long voyages. When conditions are favorable, kestrels prefer not to leave the areas in which they normally brood, and only when necessary will they temporarily withdraw to a warmer climate for the winter.

The peregrine falcon, which is distributed all over the world, is even more familiar to man than the kestrel. It has been used for hunting since time immemorial. Falconry has its own special rules, vocabulary, rituals, and its share of history: "In March," reported Marco Polo around 1290, "Kubla Khan has the habit

The Old World kestrel (*Falco tinnunculus*), which ▶ is not as fierce as it looks, does not flee man.

Opposite:
The peregrine (*Falco peregrinus*), which is not as tame as it might appear, maintains an unconquerable love for its solitary freedom, even though it has hunted with man since time immemorial.

Its long wings and tail give the kestrel excellent control over its flight. When it comes down on its prey—often a mouse—it first slows the speed of its fall with a push of its wings, then dives straight down. ▼

of leaving Cambaluc. He takes about ten thousand falcons and falconers with him."

Small birds, starlings, thrushes, swifts, larks, make up 75 percent of the prey of the lovely peregrine. But crows, ducks, and partridges are threatened as well, and the messenger pigeon population has diminished considerably because of peregrines.

But is this admirable bird, capable of attaining extraordinarily high speeds, a wanderer as its name indicates? Does it migrate? According to some, it flies south each winter, returning in the summer; according to others, only the northern species migrate.

Peregrines use the eyries of other birds, such as gray herons and ospreys, in quarries, ruins, rock hollows, or belfries, as their own ready-made nests. In April, the female lays three or four eggs. As long as she is brooding, the male brings in the food and occasionally even takes over the care of the nest and the eggs. Four weeks later, the baby birds are hatched. The father continues to be responsible for food gathering, and the mother distributes it. Like most birds of prey, the peregrine couple—which remains together for years—sends its offspring out into the world only after they have acquired the basic skills of hunting birds.

▲ The downy white peregrine young, huddled together, demonstrate their approval of the prey brought in by the father. The mother will pluck and distribute it.

The Bat

Some sixty million years ago, the first bats appeared on earth. The only flying mammal of the animal kingdom could well be our cousin since, according to some paleontologists, man and bat are descended from related primitive insectivores. But no one really seems to wish to presume on this dubious ancestry.

The bat is perhaps best known for its use of the principle of radar. Among certain species, the throat, or sometimes the nose, emits ultrasonic waves which bounce back from any object in their path, live or inanimate. The frequency of these waves varies from thirty thousand to ninety thousand vibrations per second. The human ear, which can perceive only up to twenty thousand vibrations per second, cannot hear them. The very fine hearing of the bat registers the echo, locates its source, and thus permits the bat to fly straight to its prey. The outgrowths of skin and nasal appendages which some species have help them further perfect their discernment of echoes. Not until World War II was man able to develop an equally efficient instrument for echolocation.

Bats use their hands to fly. While in birds the entire arm holds up the wing, in bats the disproportionately long fingers subtend the membrane of the wing along the entire body. The only parts of the body that are free are the head, the clawed thumbs of the front limbs, and the toes of the hind legs. In order to keep the membrane flexible, the bat regularly licks it with a smelly grease secreted by glands located between the eyes and nose.

There is no real comparison between a bird's graceful flight and the fluttering of this mammal, whose body rocks up and down with every flap of its wings. Yet it prefers heights, can move very rapidly, and its air speed compares favorably with

◀ The large mouse-eared bat (*Myotis myotis*) is characterized by long thin ears. A covering closes off the orifice of the auditory duct. The body of this bat is only three inches long, but its wingspan is sixteen inches.

Nasal appendages complete the radar system of ▲ these bats of the family Phyllostomatidae, fruit-eating bats of tropical South America.

◀ The hands act as a brace for the wings, which are actually membranes stretched taut between the long fingers. At dusk the bat may be seen fluttering about; but it can fly fast and for long periods. Its cry, in oscillating ultrasonic waves of thirty thousand to ninety thousand vibrations per second, is echoed by the objects it encounters. The ear receives the echo, and the bat thus receives directional guidance.

These large mouse-eared bats hanging from their toes spend the winter together hibernating in caves. They inhabit Europe, central and southern ▼ Asia, and Africa.

The lesser horseshoe bat (*Rhinolophus hipposiderus*) is common in the middle altitude mountains of Europe; in the east, it lives in the foothills of the Himalayas. ▶

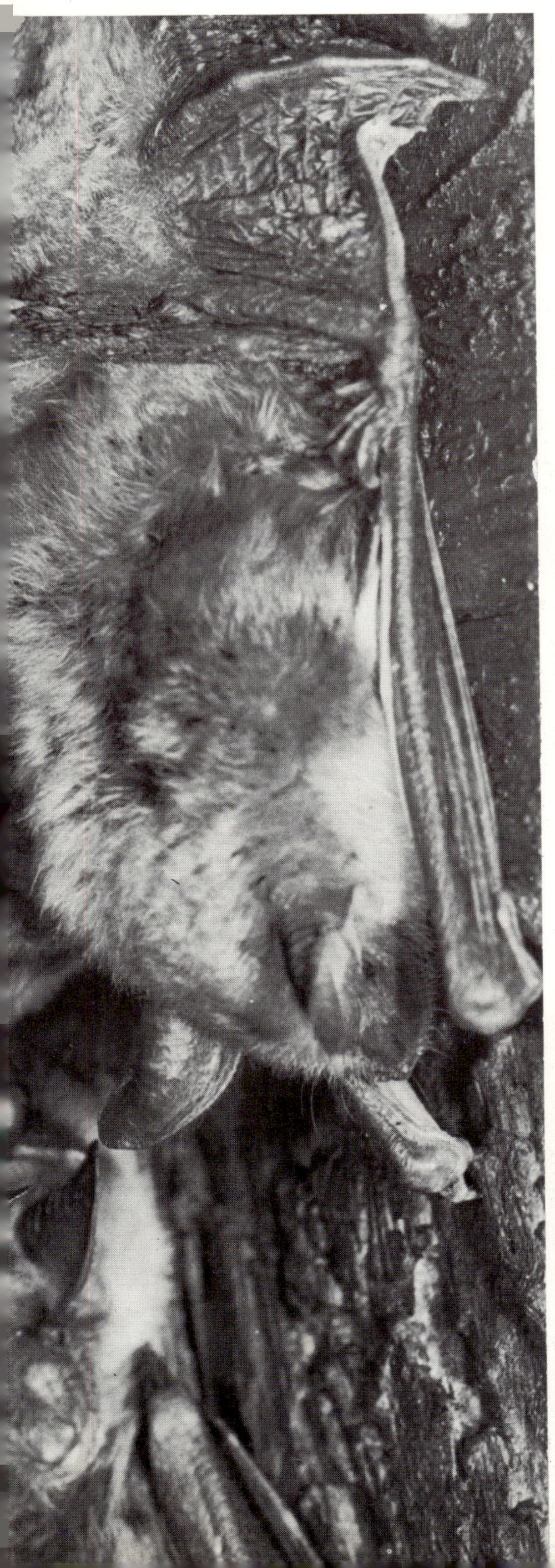

that of most birds.

The bat goes out only at dusk, when its prey, moths, beetles, and gnats, are out, too. On its initial expeditions it uses its radar to find its dark den. Afterward, it returns home by "memory," sometimes seriously hurting itself if there are new obstacles in its path.

Bats sleep head down, hanging from their toes, in some dark recess. They hibernate in groups, often by the thousands, in caves. Their metabolism slows down, the body temperature drops to 32° F. Among most European and North American species in temperate zones, mating occurs at the beginning of winter. The sperm remains in the female's uterus, and actual fertilization does not take place until the spring. A hairless offspring is born at the beginning of June. It immediately grabs the fur on the mother's chest, biting solidly into her teat, and hangs on for two weeks, even during the adults' nocturnal expeditions. At two to eight weeks, it is left in the cave, where it continues to develop. Its adult life lasts about fifteen years.

The lesser horseshoe bat wraps itself up in its wings, which now resemble a folded parachute, when it sleeps head down. Its "blanket" is nine inches wide. ▲

◀ Takeoff requires much preparation. First, the bat, which wakes up shivering, must warm up; then it begins to open its membrane; and finally it drops off, pushing off with its fingers, and begins to flutter.

Small Owls

The smaller owls are the natural enemy of many small birds. Just as the wolf fooled Little Red Riding Hood, these owls like small song birds all the better to eat them. At night, especially in winter, an owl creeps up stealthily in the neighborhood of the sleeping nests and suddenly beats its wings violently. The rudely awakened birds peep and cry in alarm, making so much noise that they reveal their exact position to their predator. However, if the birds should come upon their nocturnal attacker during the day, they take their revenge: all the birds in the area are alerted, and they corner the owl for many hours against a tree, its eyes half closed. Trying to drive it out in the open, they rush against it in order to startle it. The owl is careful not to move, but it is given a very hard time nonetheless. Actually, only 25 percent of the owl's diet consists of small birds; it prefers mice, rats, dormice, moles, frogs, lizards, and insects.

The tawny owl (*Strix aluco*), which can be twenty-eight inches long, inhabits northern Europe and northern Asia, where it settles in quarries, rocks, parks, large gardens, and old graveyards, using for its nest a hollow tree or the abandoned eyrie of a bird of prey. Its terrifying mating call, which is heard in February, begins with a sinister-sounding ululation that becomes a continuous crescendo of tongue smacking, humming, and demonic

These young owls line up on a branch and beg shamelessly for food. In order to feed their young, the parents, generally nocturnal hunters, do not hesitate to face the dangers of light and the ill temper of small song birds, which attack them in groups.

▼ The round innocent eyes of the tawny owl (*Strix aluco*) pierce the deepest night.

▲
The misunderstood little owl (*Athene noctua*) was the symbol of Athena, the Greek goddess of wisdom; according to popular superstition, it is the harbinger of death. Yet it is the most amusing of all the birds, the clown among the owls.

screams. At the end of March, the female broods the eggs—a maximum of six—for twenty-nine days, using this rest period to molt. The male feeds the mother and the young. At five weeks the young become self-sufficient. There are often two nestfuls a year.

The little owl (*Athene noctua*), which is smaller (only nine inches), lives mostly in North Africa and central Asia. It nests near man in old sheds and ruins, but it is not popular with its human neighbors since it is believed in local superstition to be a harbinger of death. Its large yellow gorgon-like face does not make its appearance any more reassuring. Yet it is a harmless and indeed useful bird, as it eats pest insects and mice, and it is irresistibly

comical to man: it has a clown's repertoire of funny faces. Its playfulness should outweigh its morbid reputation—it winks, displays its feathers, feigns illness, and seems to imitate human concentration by perplexedly scratching its head. Although a nocturnal animal, it enjoys the sun; the performance over, the little owl dreams peacefully in the heat.

▲
Just born, these young little owls will grow up to be useful adults, destroying mice and insects; still, they are persecuted because of man's unjustified distrust.

10. Marshes and Tundra

▲ The young storks must feed themselves; the parents leave the food on the edge of the nest for them.

The White Stork

When they see the V-formations of long-legged birds high in the air, ordinary men dream of distant voyages. Scientists, on the other hand, ask themselves certain questions. Although the why of these migrations is understood, most of the hows are still unknown. How do these birds orient themselves so accurately on their mass journeys? By marking individuals and systematically studying the flight patterns of white storks, it has been established that the populations living east and west of an imaginary line from Leyden in The Netherlands to Kempten in southern Germany take different routes on their migrations. To avoid the perilous crossing of the Alps and the Mediterranean, storks from eastern Europe detour over Turkey and the Middle East on their way to southern Africa; the western populations fly via Spain and Gibraltar. Knowledge of and preference for each itinerary seems to be instinctive and unchangeable within each group. Young storks taken across the demarcation line follow the route of their original group. In some cases, they will first return to their nest and they fly off in the "correct" direction.

These lovely white waders have long figured in European folklore. They build their nests on the edge of a pitched roof, on top of a belfry, or on a wooden wheel supported by a pole that man has made for them. The house or village that had "its" storks was believed to be assured of peace and prosperity. Storks were known to be faithful to their homes, and their annual return was joyfully celebrated. But stork celebrations are rare today. Civilization has chased away these good-luck birds by encroaching on their hunting grounds in damp meadows and marshes. Only Spain

▲ Any current will do for these experienced navigators; they can use the slightest air movement for gliding, wings outspread anc neck and legs horizontal, in enormous circles over their territory.

◀ The neck, with its ruffled feathers, supports the beak of this tired wader.

The white stork (*Ciconia ciconia*) meticulously grooms its young, scratching and smoothing out their feathers. ▶

and North Africa, the Balkans, White Russia, and Asia Minor still provide suitable territories for them.

Though the storks' reputation for constancy of dwelling is well deserved, they are less faithful in their mating. When a stork couple returns to their nest—a solid construction of branches, reeds, and hay, combined with rags, bits of paper, and shreds of skin—they sometimes find that a usurper male has moved in. The males then fight each other. The winner takes possession of the nest and readily chooses whichever female responds to him. His mate of the previous year then appears on the scene, and the two female rivals grapple with each other while the male imperturbably awaits the outcome of the fight, and mates with the victor.

The brooding period lasts thirty-three days. The baby storks are raised by both parents, and can fly at two months. The female generally stays near the nest, and the male seeks the food; neither parent is obliging enough actually to feed the young, but leaves the food on the edge of the nest. On the other hand, in hot weather they refresh their young by pouring water directly into their wide-open beaks; the adults transport the water in the inflatable pouch of their gizzard. If the sun is too strong, the parents use their wings as a parasol over the nest.

An excellent glider and an experienced navigator, skilled at using the slightest air current, the stork moves smoothly in the air, wings outspread and neck and legs horizontal. It is also a good walker. When it hunts frogs, snails, lizards, or mice, it moves carefully, without the slightest wing motion, and even when it wades in the marshes it never loses its dignity. In autumn, when food becomes scarce, storks form their squadrons, and one fine morning they fly away toward the sun.

Considered a bird of good omen, the stork is always welcome when it returns each year to its nest. But this event is celebrated increasingly rarely: civilization has driven the stork away by reducing its hunting territory in damp meadows and marshes. ▶

▲ The male is not a faithful partner, and changes mates with the greatest of ease.

The annual migrations of the storks follow fixed itineraries discovered by tracking marked birds. But the infallible directional sense that allows young storks on their very first trip south to find their way—even if isolated and without a guide—is still not fully understood. ▶

Baby storks tend to become dehydrated in hot weather. Their parents refresh them by pouring water, which they carry in the inflatable pouch in their gullet, into their beaks. ▼

Reindeer and Caribou

If the great herds were suddenly to disappear from the Arctic, the human life there would be wiped out as well: the reindeer is the sole means of survival of the inhabitants of the far north, whether Finns or Lapps, Eskimos or Indians of northern North America. The reindeer furnishes milk and meat; its skin is used for blankets, shoes, clothing, and tents; its

◀ Scouts lead the way for these giant herds of Old World reindeer (*Rangifer tarandus*). Moving along side by side, in the spring they head for the northern meadows along the fjords; in autumn they return to the thick southern forests.

Their large cloven hooves acting as snowshoes, these deer of the far north can walk in the muddy marshes or thick snow of the tundra without sinking. ▼

sinews for rope; its bones and antlers for tools. Finally, it is a beast of burden for children, the old, and the sick. The reindeer is even more indispensable on the tundra than the camel is on the Sahara: the reindeer is the sailing ship of the icy deserts.

The reindeer's history is almost as old as our own. At the end of the Ice Age, it

◀ New World caribou are totally wild; these European reindeer live in semifreedom and have never allowed themselves to be completely domesticated. Each year, Laplanders gather the herds in giant enclosures in order to brand the young and cut out the animals to be slaughtered.

was scattered over a large part of Europe, as far south as the Mediterranean. The oldest traces discovered date back three or four hundred thousand years. Reindeer are commonly represented in the frescoes of prehistoric caves in France and Spain. The discovery of many bones show that during the Late Paleolithic Age it was the favorite game animal of early hunters. Man began to attempt to tame and raise reindeer very early; in A.D. 9, a Viking, Ottar of Helgoland, boasted of owning a herd of six hundred head.

Today the reindeer's twenty subspecies live in the tundra of the Old and New Worlds; in the New World reindeer are generally known as caribou. The best known subspecies inhabits Scandinavia, Lapland, and eastern Europe. It measures almost eighty inches and is approximately three feet wide at the withers. Its long, brittle and delicate fur is dark brown in summer and lighter in winter. Its legs are quite dark, and its neck and throat are white, as are the manes of the old males. Both sexes have permanent antlers, but those of the female are larger.

Of all the deer, reindeer are the most sociable. They gather spontaneously in great herds and move along companionably, always following experienced males who serve as scouts. When the herd is at rest, it is the old females who serve as the lookouts. Reindeer feed on lichen, mushroom, young plants, and kelp, as well as on lemmings, mice, birds, or eggs. The mating period occurs at the end of autumn and lasts about three weeks. Gestation ranges from 216 to 246 days.

Forest reindeer, known in the New World as woodland caribou, do not migrate. Reindeer of the tundra do, and herders follow them at their own risk. This semiwild deer has never allowed itself to be completely domesticated, and nothing can confine it. In autumn the herds move south to the forests, returning north in spring to the meadowland near fjords, avoiding the Arctic swamps, where swarms of mosquitoes make summer miserable. Unable to control the herds, the herders follow them, perforce becoming nomads. In Lapland, about two hundred thousand reindeer are "owned" in this half-domesticated fashion. They are gathered together several times a year to brand the young and cut out those to be slaughtered. The rest of the time, the reindeer are left free. They do not need man; it is man who needs them.

Neither rapids nor icy waters can stop the herds: reindeer swim remarkably well. Their thick, air-filled fur protects them from cold and humidity. They feed mostly on moss, mushrooms, lichen, kelp, and leafage.

The Common Crane

▲
The always dignified common crane (*Grus grus*) moves slowly in the marshes on its long wader's legs. It stretches its neck straight forward when in flight and keeps its legs stiff in the same horizontal line.

Above:
The golden-feathered crowned crane (*Balearica pavonina*) inhabits tropical Africa. It is approximately three feet long.

The marshes and peatbogs once resounded with the loud blare of trumpets that could be heard for miles around: the cranes were giving their mating call. But where are their songs now? In draining the marshes, man has chased away the cranes; today their call can only be heard high overhead, as they fly from Scandinavia to Ethiopia or India, where they spend the winter.

The cranes' great mating ritual, one of the strangest and most beautiful in the animal kingdom, occurs in the spring. Smolik describes it this way: "As if imbued with a sense of the deep significance of this moment, the male, attracted by the female's trumpet call, stands in front of her. The two proud birds face each other, immobile, taking each other's measure, for a long time. Then the male raises his wings as if to embrace her, but the more reserved female moves away. He follows her in long solemn steps, holding his head high and his wings closed. There is another eye-to-eye appraisal. The male then bows and scrapes in all directions; he performs graceful dance steps, gravely and precisely executes backward loops, sways his hips, swings around, turns in circles, agitates his rustling wings, and jumps up and down, punctuating the entire ballet with trumpetings, as if providing music for his own dance. From time to time he plucks stems and tufts of grass and presents them to his

◀Common cranes, their numbers diminishing as the marshes are drained for cultivation, brood only rarely now in Europe.

chosen one, juggling with materials for the nest as he pursues his courtship dance. The female remains modest for a long time, and mating occurs only in the secrecy of protective thickets." This ritual marks the beginning of a true marriage; the couple will never separate.

The pair builds its nest in hard-to-reach spots in marshes, in shifting soil, and even in the water. The nest is a flat trough anchored to a solid raft of reeds, in which the female lays two to four spotted dark brown eggs. She broods them for most of the time, the male taking over at set intervals for short periods. Each shift is marked by an exchange of loud cries and symbolic gestures of nest-building. After the month the babies hatch, already able to swim. At eight weeks they attempt to fly, and feed on plants and insects like their parents. The parents protect and guide them for two to two and a half months. When adult, they will reach a length of forty-five inches.

The Moose

Andy Clarke

All ages and civilizations have their hunter's tales. In the *Gallic Wars*, Caesar wrote of the *alces*, the ancestors of today's moose, which he saw in the Hyrcanian Forest: "The appearance and color of their fur is that of a goat, but they are much larger, have no horns, and their legs are jointless. This is why they cannot lie down to sleep and must lean on trees. If they

▲ Tranquil, solitary, self-sufficient, the moose (*Alces alces*) roams the ancient marshes of its territory. It needs unlimited space; each animal occupies a territory of two square miles.

happen to fall, they cannot get up again. When hunters identify which trees they customarily lean on, they dig under the roots or saw into the trunk. Now when the animals lean on them, the trees will fall, and the moose fall, too."

Aside from the peculiar hunting technique reported by Caesar, this description by a modern zoologist, Floericke, is cer-

Moose have no continuing family sense. The female raises her calf alone, and as soon as it can, the offspring leaves her to live alone.

tainly in the same tone: "A coarsely shaped horse's skull, a monstrous spur-shaped muzzle, tiny, shifty, blinking, porcine eyes, unsightly long-whiskered dewlaps, heavy spatula-like antlers, a camel's hump on the front quarters, the stiff and abrupt hindquarters of a giraffe, the shiny, long, white legs of a wader, a ridiculous stub of a tail . . . all of this in a single strange giant animal."

There were moose as early as the Ice Age, when they roamed Europe as far south as northern Italy, and North America as far south as New Jersey. They have retained the powerful and fascinating ugliness of survivors of a previous age and another world. Floericke's "strange giant animal"—it can be as long as twelve feet, stand seventy-eight inches at the withers, and weigh 1,100 pounds—presently inhabits only Norway, Sweden, Finland, the Baltic countries, Poland, Russia as far as the Yenissey River and Lake Baikal, and some areas of Asia and much of the north woods of North America. World War II almost decimated the European moose population: the 1,400 animals living in the area of the Gulf of Riga were wiped out. Only a few living in the swampy regions of Poland survived, by taking refuge in the giant marshes of the Russo-Polish border. To guarantee their survival it was necessary to create new reserves of stock. In 1951, the Poles settled a moose family consisting of father, mother, and two calves in the three hundred square miles of the Cerwone Bagno. In 1959, 180 head were counted by helicopter. The Swedes used protective measures to increase their moose population from thirty thousand in 1939 to fifty thousand in 1962. According to their experts, these great cervines are steadily spreading north.

The one absolute need of the moose is space: each animal requires a territory of two square miles. It moves in springy strides with a sure gait on its stocky legs that end in deeply cloven hooves, and thus never sinks in the unstable ground of the marshes or in sand or deep snow. It swims lakes, streams, even sea inlets, and easily clears obstacles over six and a half feet high. Considering its bulk, it slips through undergrowth with amazing lightness. It

▲ The calf is born sleek and unspotted in May or June after a gestation period of at least eight months. It is nursed by its mother until the following mating season.

◀ Moose enjoy clear water, and will bathe in ponds, lakes, or shallow ocean inlets.

▲
Phlegmatic and solitary eleven months of the year, the male becomes ferocious in September during the mating season. Then it obstinately follows females, viciously fighting all rivals.

grasps food with its large, mobile lips, eating leafage, conifer shoots, huckleberries, briars, and, in winter, tree bark. The adults, who have difficulty reaching the ground, prefer to graze on trees and bushes. The young kneel and graze at ground level. Moose feed mostly at dusk or at night, and constantly change their daytime resting places. They will sometimes lie down to rest; more often, however, they doze standing up.

Mating, however, takes place in a particular spot dug in the forest soil. During the mating season, the phlegmatic personality of the moose changes: in September, the male feverishly plunges through the swamps and sandy moors, trots about the fields and meadows, even venturing into inhabited areas in pursuit of a female whose tracks he is following. If he meets a rival, they fight furiously, using hooves and antlers.

After a gestation of 240 to 260 days, the young cow gives birth to one calf in her first litter; subsequently she will bear two or, very rarely, three. She nurses them until the following mating season, and then becomes solitary once more; as soon as it is able, the calf leaves the mother of its own accord to live alone.

The cows, which are almost as large and heavy as the males, have no antlers. In the males, the antlers first appear in the second year as points. In the third year the branching forks appear, and in the fourth year the first back tines. Then the final palmate structure develops, shaped like a large triangular and jagged spade. The antlers now have a span that may be three feet across.

A young moose at the end of summer; note that the velvet is gone, rubbed off against branches and rough surfaces. ▲

The back tines of the second head crown the four-year-old moose. At this time of year the antlers are still covered with velvet. ▶

The Wolf

Through the ages an aura of terror has surrounded the wolf. Man has long recognized this animal's ferocity and bloodthirstiness, its courage and its wild love of freedom. But it is from this same savage creature that the dog, man's faithful companion, developed. And it is true that a wolf cub captured at birth can be tamed and domesticated like a dog. How to reconcile, how to explain this paradox? No one yet has. Even in ancient legend, the stories of children tenderly raised by

◀ The timber wolf (*Canis lupus*), ferocious in the wild, can be as tame as its descendant, the dog, when it is raised in captivity. When wolves are crossed with dogs, the resulting hybrids are sterile.

The wolf lives all its life in a family group. Couples are monogamous and faithful, and the offspring of their successive litters remain with them. When the parents want to hunt together, an older female watches the cubs. ▶

wolves—the best known, of course, is that of Romulus and Remus—are almost as common as tales of the evil devourers of herds, their keepers, and Little Red Riding Hoods.

The European wolf, which is forty-five inches long and stands thirty-five inches at the withers, is quite different in appearance from the wolves of the far north, Asia, and North America, but is of the same species; the habits and behavior of these various populations are identical. All are nomadic hunters, prowling deserted areas in the afternoon and moving into inhabited regions at dusk. Always hungry, they indiscriminately kill all the prey they can find—weak and sick large game, domestic animals, foxes, hedgehogs, birds, reptiles, and mice—and feed on plants as well. When a wolf encounters a herd of domestic sheep, carnage ensues: the wolf is seized by a frenzy and tears apart animal after animal, far beyond its immediate needs; this, of course, is why the wolf is so hated and feared by shepherds.

In spring and summer, these carnivores hunt alone or in small family groups. In autumn, they form packs, and by their numbers can then attack large game that they could not overcome individually. The pack, led by a strong male, is governed by a strict hierarchy, and there is a constant division of labor: one group of wolves frightens the prey—stag, moose, reindeer, horse, sheep, or cow—while another cuts off its retreat, and a third pursues and kills the victim. When the winter is long and hunger too piercing, wolves forget all caution and invade built-up areas, where they try to force in cattle sheds.

Wolves have an exemplary sense of family. Couples remain faithful for years, and gather in tightly knit tribes led by the oldest male that include different generations of offspring. During the mating season in January or February wolves do not fight as violently over the females as one might expect. The males do fight, but the contest seems to be simply a means of establishing dominance; the winner never bites the throat offered to him by the loser, and in no way takes advantage of this gesture of defeat.

The female wolf gives birth to four to six cubs, sometimes in a badger's burrow or a fox hole, sometimes in a lair that she builds herself. She nurses and cares for her young, and when they are old enough to be weaned the father brings them meat, which the mother premasticates and distributes. When the cubs are older they eat carcasses that the parents bring them—this is the cause of the odor that often emanates from a den. Wolves are sexually mature at age two. They live about twelve more years, always ferocious, always noble, and always in their close family groups.

The Bittern

The bittern's cry in the marshes at night sounds like the last scream of a drowning man. Yet this chilling call is actually a love song. The rattling, gurgling, frantic gasps of the male bittern are meant to seduce the females—as many as possible, since the bittern is polygamous.

In each messy nest of reeds, begun by the male and completed by one of his mates, lie two to seven eggs. The female broods and feeds her family alone, while the male scurries back and forth among his harem. At one month, the babies leave the nest; at eight months they can fly. Bitterns, scattered over Europe, Asia, and southern Africa, live in the rushes and reeds in areas of alluvial deposits. They feed on water insects, snails, shellfish, and fish. When threatened, they stretch their necks, pointing head and beak directly upward and thus making their body shape as thin as possible. They defend themselves when attacked by hurling themselves at their adversary's eyes. Many of their habits are secretive, which makes observation in the wild difficult.

One variety, the little bittern, is just as circumspect as the bittern but less fearful. At fourteen inches it is half the size of its close relative. It lives in Africa, New Guinea, Asia, and Australia, where the females build communal villages on riverbanks and in reed copses. The male does his share of the construction by supplying the basic building materials. The parents take turns brooding the five to seven eggs; then together they feed the young, who hatch after approximately seventeen days. The young leave the nest for short periods after a week, moving out permanently after a month. During the mating season, the female responds to her future mate's serenade by cawing in a friendly way. The bittern is a nonmigratory bird; the little bittern flies to Africa for the winter.

Above:
Remaining unseen is often the best defense. The bittern (*Botaurus stellarus*) manages this by extending its neck and pointing its head and beak directly upward, making its body shape as inconspicuous as possible.

◀ Little bitterns (*Ixobrychus minutus*) like to live over water, and build their nests of reeds.

Hiding is almost a way of life for bitterns. They live isolated and fearful in the reeds in areas of alluvial deposits. ▶

Lapwings and Plovers

The lapwing, an Old World species, is a skilled acrobat, and his talents are best displayed in the mating season. The male announces his performance by uttering his funny cry, *Kui-vitt, Kui-vitt.* Then the show: he begins in the air, flying in loops, rolls, and nosedives to impress his chosen one. He continues his act on the ground, strutting back and forth, bowing, turning, and displaying his tail feathers. He convinces the female of the seriousness of his intentions by gathering twigs to build the nest, punctuating all of this with gestures and mimicry. The female cannot resist him. But as soon as they have mated she discovers that she is only one of her partner's many conquests. Her seducer abandons her, and she will have a nestful of young to raise alone; however, in case of danger her mate will return to defend his territory.

At the end of March, four pear-shaped eggs lie in the nest, each pointing to the center. After twenty-five days the young hatch. They can fly by June, but the ground is usually too dry for them to find sufficient food. They move to the neighborhood of a riverbank, where they spend the summer. In winter, they move again, to the warmer regions of Europe. In February they return to their birthplace, where they mate and brood; lapwings are semimigratory birds.

Plovers, which are related to lapwings, live in the marshes and peatbogs of both the Old and New Worlds. When they find a good spot to build their nest, a very simple construction, in an area providing sufficient worms, snails, grubs, and insects, they happily establish themselves, and can soon be seen trotting about so rapidly that they look like rolling balls. They stop occasionally to inspect their surroundings with their large eyes, then feverishly return to their occupation. Among plovers, both parents brood the eggs together. The young leave the nest for the first time—in their parents' care—on their second day of life.

European plovers spend the winter in Africa; those of the New World migrate to South America. The population is summoned by melodious calls, and flies off in rustling swarms.

◀ The ringed plover (*Charadrius hiaticula*) lives on sandy and muddy riverbanks and nests on beaches, between the dunes.

The lapwing (*Vanellus vanellus*) broods in marshes or damp grasslands. It spends the summer along riverbanks, leaving to winter in southern Europe. ▶

11. Rivers and Oceans

The Old World Otter

◀ The Old World otter (*Lutra lutra*) hunts for prey at night as well as during the day. It is an expert swimmer and diver, moving soundlessly in the water.

In eastern Europe, the word for otter is used as a tender pet name—perhaps double-edged, since the behavior of this aquatic mammal is not particularly tender.

The otter, which measures thirty-two to thirty-six inches with a tail half as long, has a light-colored belly and a sumptuous thick, smooth, dark coat. It is as comfortable in water as on land. On its nocturnal land expeditions, it glides along as rapidly and silently as a snake. Its body is slender, its head pug-nosed with prominent eyes and short round ears. It kills birds, small mammals, and frogs in complete silence. In the water, it stops up its nostrils and dives and swims ably with the help of its webbed paws. It can swim even under ice, providing there are holes that allow it to surface for air. It unerringly finds the breeding pools of fish hatcheries and, like a wolf among a flock of sheep, kills as many as possible, far more than it needs.

The Old World otter lives in the tree-shaded rivers, lakes, and pools of Europe and Asia, with the exception of Indochina, and is found in northeastern Africa as well. Its burrow, dug in riverbanks, exits in the river bottom twenty inches from the water line. A passage leads to a funnel-shaped chamber with an insulating grass roof; this is the "nursery."

After the twice-yearly matings at the end of winter and summer, the female gives birth in the nursery to two to four young, which she tends for six months. The young are born blind and furless, but soon become resourceful, adventurous, and very boisterous. They rush down the muddy hills in bands, like children at play, whistling and quarreling. In winter, the ice provides an even more delightful surface for their tobogganing.

The young otters gradually learn to hunt, catching fish by cornering them with their tail in hollows in the riverbank and then seizing them. Underwater they use their sensitive long whiskers to pick up any ripples made by passing fish even some distance away.

Otters must also learn how to avoid man. Only stringent legal measures protect them from man's lust for their fur. In Germany, there were only two hundred otters left in 1965.

Like children at play, young otters go tobogganing, rushing down the icy or muddy banks and diving into the water with a loud splash. ▶

On the lookout, this handsome carnivore stands up to its full height to inspect its surroundings carefully when it is on its land expeditions. When it searches for fish, it seeks out well-stocked ▼ waters and fish hatcheries, which it raids.

The Duck-Billed Platypus

Poisonous like a snake, egg-laying like a bird, a burrowing and swimming mammal like a beaver: the unique duck-billed platypus developed in its present form millions of years ago in the period when the mammals were just beginning to become differentiated from the reptiles. It lives only in eastern Australia and in Tasmania, in quiet waters as well as running rivers in the shade of forests.

Its heavy cylindrical trunk, which varies in length from fourteen to eighteen inches, and sometimes, according to some naturalists, is as long as two feet, stands on short hind legs. Its paws are webbed and clawed. A coat of shiny bristles covers a short woolly fleece; the flat five-inch tail has bristles only on top. Two tiny eyes rise above a long horny beak, and a flap of skin seals the ear openings when the animal swims or dives. The male has a corneous spur on the heel of each foot that connects to glands in the legs which secrete a poisonous substance. The platypus can cause serious wounds with this structure.

The duck-billed platypus cannot consume large prey. Like the duck, it digs underwater in the mud with its beak, which acts as a strainer to trap its food. Its catch—insects, worms, crabs, and larvae—is mixed with saliva in the cheeks and then chewed. The platypus is a good swimmer and diver, but cannot go for more than two minutes without breathing and must surface regularly.

Like the beaver, the platypus uses its long solid claws to dig burrows in riverbanks. Fortified corridors often exiting underwater lead to a spacious, ventilated "living room," where the duck-billed platypus stays during the day, usually with a mate. The female sets up a "nursery," a hole filled with hay and eucalyptus leaves, where she lays two or three half-inch-long eggs, which she broods for fourteen days.

At three days, the young are just over an inch long, and at seven weeks they are two inches long. To nurse, they gently tap their mother's abdomen with their beak, which is still white, and lick the drops of milk secreted by specialized sweat glands: the mother has no teats and must lie on her back to feed her young. At five months, the duck-billed platypus becomes self-sufficient. For ten or fifteen more years it will live the strange existence of a relic—a living reminder of the days of the early dinosaurs.

A survivor of an age millions of years past, the duck-billed platypus (*Ornithorhynchus anatinus*) is the only mammal that lays eggs and is poisonous. It catches its prey by digging with its beak in the mud bottoms of rivers and inlets. It uses its burrowing claws to tunnel in the riverbanks. The duck-billed platypus is found only in Australia and Tasmania.

The Beaver

When some years ago a group of citizens, tired of waiting for an administrative solution to a housing crisis in France, decided to build their own homes, they called themselves "The Beavers." The sophisticated architecture of that animal is one of the natural wonders of the world: not only does it build extremely complex homes, but it changes the topography of its territory, alters the course of rivers, and builds dikes and dams worthy of experienced human urban engineers.

As soon as it finds suitable terrain, a beaver couple begins to tunnel winding passages several yards long along the high riverbanks. These corridors end in spacious round "living quarters," which are always above water level, preferably in the roots of a tree. When this is done, the beavers start building a second residence on dry land, which has at least one underwater emergency exit. They first dig deep trough-shaped foundations, and then build walls, using the excavated dirt. They build a vaulted roof of branches and sticks, cemented with dirt, mud, moss, or grass. Then they must guarantee a regular and controlled flow of water to provide a swimming entrance to their dwelling. They divert the flow of the river or stream by building dikes of logs and branches reinforced with rocks and cemented with mud; these can be as long as 325 feet and 10 feet high.

Beavers mate in February or March. After a gestation of 205 days, the female gives birth to two to five young, which can already see and hear. They are not weaned for a year, sometimes even longer. At four they are adults, and measure thirty to forty inches long, twelve inches high; they weigh forty-five to eighty pounds. They are clumsy on land, but swim gracefully and rapidly, using their scaled tail as a rudder and their webbed paws as oars. They are strictly herbivorous.

Though the beaver's achievements have always fascinated man, he also has been interested in the animal's fur and meat—its tail was considered "lean" and therefore was permitted during Lent. In the Middle Ages, a musky secretion of the beaver was used to treat convulsions; this substance is still sold today in the Middle East for a high price. Beavers were once on the brink of extinction; today they are protected by strict laws.

◀ The beaver, lumberjack, engineer, urban planner, and architect, flanks its underground fortresses with log superstructures. It fells and trims trees using only its natural tools: its paws and teeth.

Beavers (*Castor*) are protected by thick fur, and ▶ can face the snow and ice in their winter searches for food.

Gnawing on the trunk, this small lumberjack seated at the foot of the tree is attacking it eight inches from the ground.
▼

The beaver, a championship diver, can remain underwater for fifteen minutes without coming up for air. Stopping up its nose and ears, it uses its webbed hind paws to row and keeps its front paws tightly at its sides. ▼

Crocodiles

Over a million centuries ago, giant crocodiles ruled the earth. Crocodiles already existed in the Cretaceous Period, which began 140 million years ago and ended 70 million years later when great parts of the present-day continents were still under the seas and the Alps and Rockies were beginning to form. Crocodiles spread over all the land and waters. Fossils of their ancestor *Protosuchus* have been found in rocks of the Triassic Period, at the beginning of the Mesozoic Era.

Crocodiles are undoubtedly the most developed of all the reptiles. They have a four-chambered heart and teeth in sockets, which are constantly replaced as they wear down. This high level of development and the animal's relative security from predators explain how crocodiles have been able to survive until the present day undergoing practically no evolutionary changes.

Crocodiles live in slow-moving streams, lakes, marshes, and coastal lagoons. They slip through the water like a torpedo, limbs outstretched, propelled by undulations of the giant tail. When they dive or swim underwater, valves seal the ear openings and nostrils; the mouth can remain open, since another valve in the back of the throat prevents water from entering the respiratory passages and the digestive tube. When cruising in the water, only the crocodile's two eyes and elevated nostrils, now open, show. Air entering the nostrils moves through nasal passages above the roof of the mouth directly to the trachea, bypassing the oral cavity and thus permitting the animal to breathe and eat at the same time.

During the day, crocodiles doze on riverbanks, sandbanks, or floating tree trunks, warming themselves in the sun. At night, they sometimes go on short land expeditions, more out of curiosity than for food. Their diet consists of fish, birds, and animals that come to the water's edge to drink. These are rapidly seized and dragged into the water, where they are ripped apart and eaten.

The female buries her eggs — twenty to thirty of them — in the sand, or hides them in a hillock of fermenting plants. She keeps careful watch over their nesting place. After nine or ten weeks, the young cheep for their mother to come help them break out of their eggs. When they hatch, the young are twelve inches long; they can already run, and they immediately head for the water. If they manage to escape being eaten by the old crocodiles, they can live about fifty years.

▲ Its mouth open as it swims, the crocodile breathes through its nostrils. A valve in the throat keeps water from entering the respiratory passages and digestive tract.

The crocodile's apparently clumsy gait on land is deceptive; it is capable of moving surprisingly ▼ fast.

Crocodiles, which have survived for so many centuries without natural enemies, have finally themselves become a prey animal: they are hunted by man for their skin. It has become necessary to impose strict measures to protect them.

The terrifying toothy Nile crocodile, which is as feared among animals as it is among men, does have one friend: the crocodile bird. Eight inches long, this bird calmly walks into the mouth of the crocodile, picking out and feeding on leftover food and leeches attached to the frightening jaws. In case of danger, the crocodile bird chirps loudly, and its "ward" gratefully submerges.

Even a dangerous saurian can have a friend: the ▶ crocodile bird dares to enter the crocodile's wide-open jaws to search for food, and emerges safely. In exchange, the bird warns its host of impending danger.

A view of the inside of the mouth of the Nile crocodile (*Crocodylus niloticus*) should be ▼ sufficient warning.

The Fire-Bellied Toad

Scottish and Breton legends tell of underwater ghost towns, and it is said that when there is a full moon one can hear their bells toll. The same dim melancholy tolling that seems to come from the very depths of the sleeping waters can be heard when the fire-bellied toad sings its song.

The fire-bellied toad remains faithful to the neighborhood of its birthplace—a grassy pond, a shallow pool, a water-filled ditch—all its life. During the spring, summer, and autumn, the toad stays in the water, with only its eyes and nose above the surface, enjoying as the heat of the sun. It is timid and cautious, and at the least hint of danger it dives down in a flash and sinks into the mud. It waits for the danger to pass, and then returns silently and cautiously to its original spot. It swims as easily as it breathes; it hunts for food—insects, worms, and snails—at night.

At the end of autumn, the fire-bellied toad must find a place to hibernate. It moves agilely onto land, leaping and bounding, and seeks out a spot where it can bury itself deeply enough to escape the frost. If en route it faces some danger it cannot escape, it lies on its back feigning death, exposing its brightly colored stomach, and its skin glands exude a corrosive foam with a very bitter odor. The enemy, frightened and repelled by this, will often give up the attack.

Fire-bellied toads spawn several times during the spring months. The eggs, which are not numerous, are deposited on aquatic plants or slipped to the pond bottom. The tadpoles hatch in a week, and grow to a length of two inches. By August or September, after the final metamorphosis, they have shrunk: the young toad adult is only three-fifths of an inch long. Only the last-born of the season are not ready to move on to land and hibernate; these remain tadpoles all winter. Fire-bellied toads are sexually mature in their third year.

The fire-bellied toad is a species of the plains of northeastern Europe, and is found as far as the Urals and northwestern Asia Minor. The male of this species can blow up a vocal sac like a balloon; this is what produces his characteristic tolling sound. Another species, a close mountain relative of the middle altitude lakes of western and southern Europe, is the variegated fire toad; its belly is yellow rather than crimson.

The soft April sun is the signal for fire-bellied toads (*Bombina bombina*) to mate. The eggs are deposited on aquatic plants or slipped to the pond bottom. The development of the tadpole is complete by the beginning of autumn.
▼

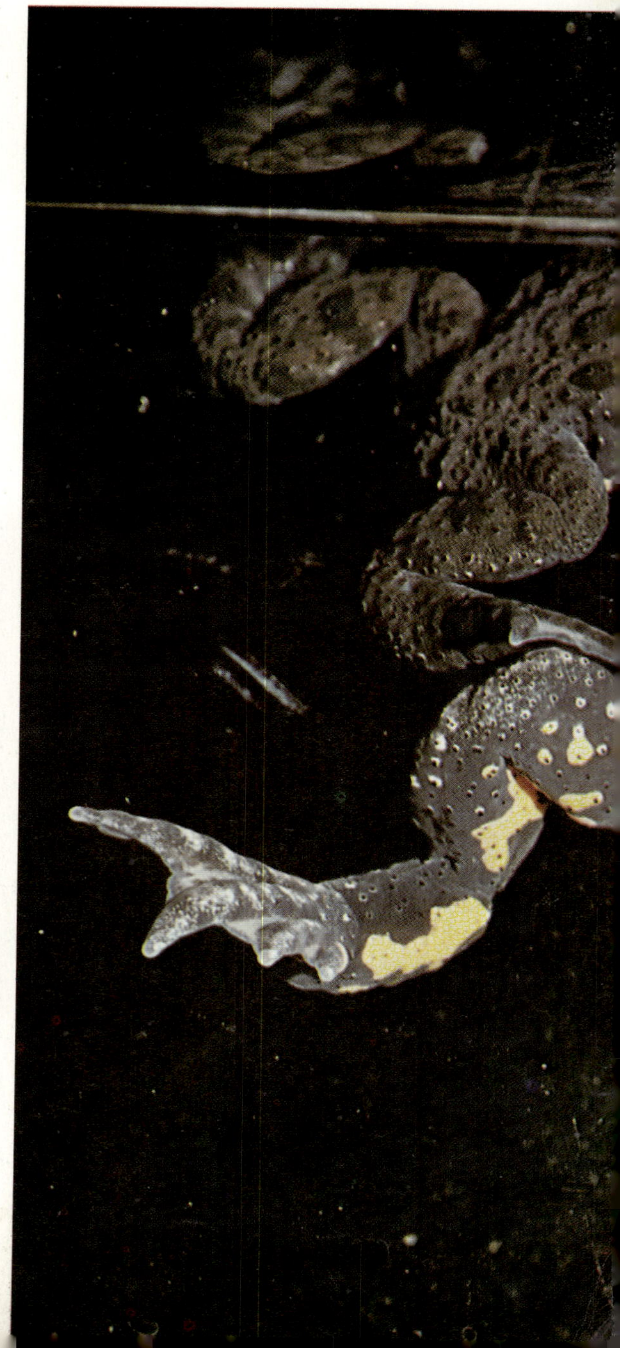

A melancholy tolling like that of the bells of some mysterious underwater city heralds the mating season of the variegated fire toad (*Bombina variegata*). ▶

Laziness, comfort, and security induce the hippopotamus (*Hippopotamus amphibius*) to remain in mud or shallow waters. Thus almost submerged in its environment, it can indulge in its favorite pastime of doing nothing. A species of egret warns the hippopotamus of danger and rids it of parasites. *Opposite, below:* The appealing, awkward baby never leaves its mother's side.

The Hippopotamus

Large, clumsy, peaceful, and sociable, the hippopotamus lives in the marshes, streams, and shallow lakes of Africa, from Senegal and the Sudan to Natal. The male is more than fourteen and a half feet long, stands five feet high, and weighs two or three tons. Its large head has eyes, nostrils, and ears that look ridiculously small in relation to its overall size. It is perhaps an ugly animal, but even its canines, which have evolved into tusks weighing four and a half to six and a half pounds, twenty-eight inches long, cannot make it look very mean. It is a greedy herbivore, consuming about sixty-five pounds of plants a day.

The hippopotamus is clumsy on land, since it is off balance because of its great weight and short legs; it is really only at ease in the buoyancy of the water. It swims tirelessly, diving deeply, and is capable of holding its breath up to six minutes at a time. Mating, birth, and the nursing of each offspring take place in the water. The behavior of mother and child together is moving and tender; they never separate. The female is dangerous only when her

young is attacked. Family life, in its narrow sense, is unusual among hippopotamuses, but mothers and young do gather in herds sometimes as large as fifty.

Though fond of company, the males are extremely jealous of their own territories, which they establish by spreading their excrement on the ground with a swirl of the tail. Rivals must not enter the area demarcated in this way. Hierarchy of dominance among hippopotamuses is similarly established: two contesting males spray each other with loose excrement, and the loser is he who gives up first. This "combat" wounds only the loser's dignity; the same is not true of mating battles, in which the tusks are used.

Good marksmanship is necessary in hippopotamus hunting; one must aim at vulnerable spots, as otherwise the bullets will ricochet off the armored skin. Thus, to kill a hippo was long regarded as one of the hunter's highest achievements, and in the nineteenth century, many were killed off. Today, fortunately, the camera often replaces the rifle on safari. There is as much personal glory in skilled photography—and perhaps a better appreciation of wildlife; and the hippos, who harm no one, are no longer slaughtered for man's pride.

▲ Hippopotamuses, which move like bulldozers, gather in herds to search for food and wade and wallow in the mud.

◀ Bathing involves splashes and games, regardless of age.

A friendly nap naturally follows the bath. But a mother is always conscious of her responsibilities, and never completely closes her eyes. She is dangerous when her offspring is threatened. ▶

Turtles

Here the contrast in size between a very young giant tortoise and an adult is vividly illustrated.

Turtles are a totally unique group of animals, practicing passive resistance. They have survived successfully from the earliest days of the dinosaurs. During the Age of Reptiles there were sea turtles twelve feet long and landgoing tortoises of almost that dimension.

The anatomy of the turtle puts it in a reptilian group all its own. The shell, which consists of top and bottom halves, is the animal's sole protection. It is made up of bony plates overlapped by large horny scales; the top half is curved, the bottom half is flat. Top and bottom shells are connected by two bony bridges between the legs. Only the legs, tail, and wrinkled neck ever emerge from the protection of the shell. In case of danger, the turtle draws in these parts, and horny scales protect what still shows of the limbs. There are some species, like the box turtles of North America, which have a hinge mechanism on their bottom shell that permits them to enclose their bodies completely.

Over the eons turtles have adapted to land and water, both fresh and salt, and they can be found in grasslands, forests, and deserts. Sea turtles are totally adapted to a life in the open waters. When it comes time to reproduce, however, they must come onto land, struggling painfully on their flippers, to lay their eggs. The female hides them carefully in troughs that she digs on a beach, just above the high-water mark. She tries to erase all traces of the hiding places of her future offspring, but she can do nothing about the track she

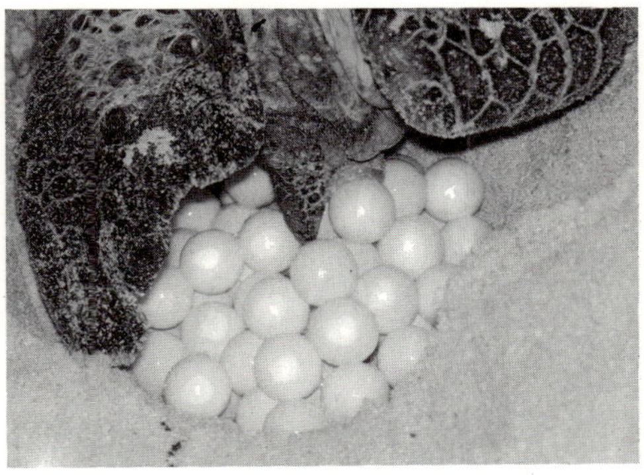

The journey from the sea to the beach where this sea turtle lays its eggs in a nest dug above the high-water mark is long and arduous.

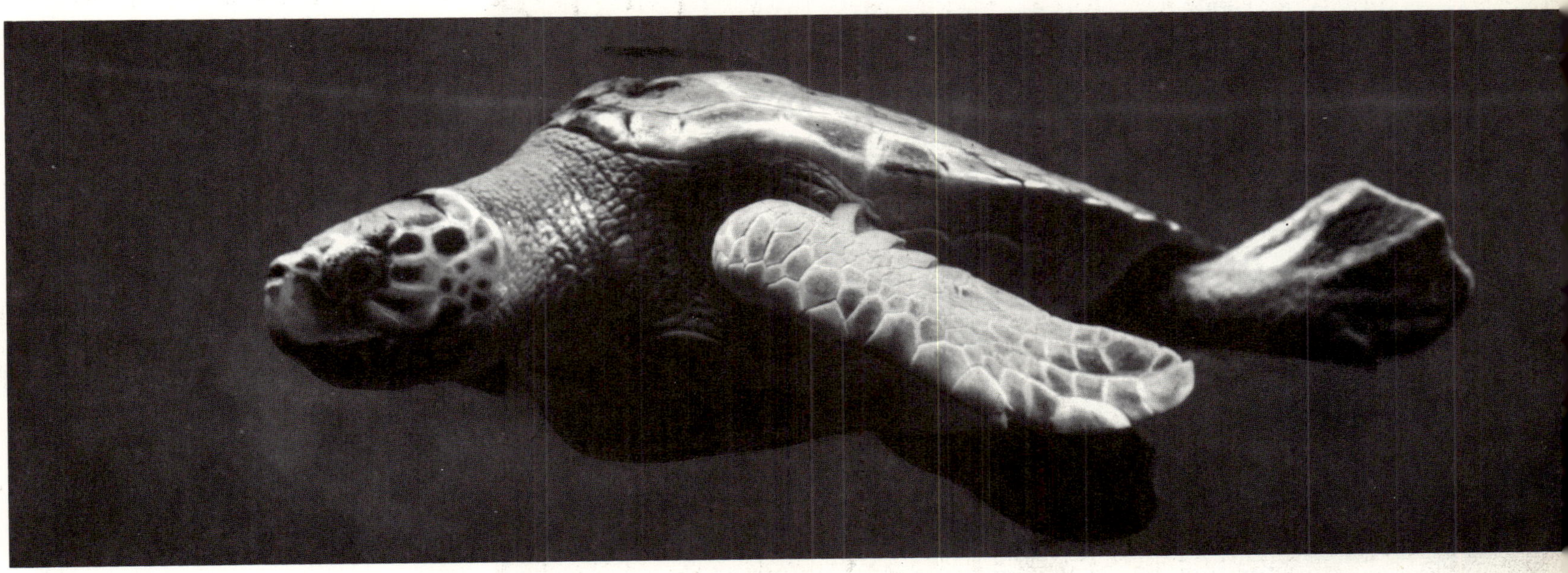

makes in the sand as she returns heavily to the sea. Thieves—human or animal—need only to follow this track to find the eggs. If the young do manage to hatch, they immediately face another danger: the shore birds hat lie in wait for them all along their slow, painful path to the sea. Only in the water can the sea turtles live in peace.

The giant tortoises of the Seychelle Islands have remained substantially as they were sixty million years ago; they are three feet high and weigh 440 pounds. Some live well over a hundred years, the longest lifespan in the animal kingdom. Unfortunately, today even the strictest protective measures are just barely keeping them from extinction.

The return to the sea marks the end of the mother's ordeal. But she cannot erase her track in the sand, and pillagers—both animal and human—follow her trail to the nest. Thus the offspring are in danger even before they hatch.

The Sea Horse

Tender embraces and languorous dancing precede the mating of the sea horses (*Hippocampus guttalatus*).

In almost all animals, it is the female that incubates the eggs. The most notable exception to this rule is the sea horse.

After mating, which is preceded by dancing and tender embraces, a female sinks her ovipositor into the male's brooding pouch and lays her eggs in it. Two or three other females follow suit, until the pouch is full—it can hold as many as five hundred eggs—and closes. The larvae, fed on nutritive elements carried in the paternal bloodstream, develop gradually, while the pouch swells at the same rate. After six to eight weeks, hundreds of young sea horses are "born" from the father's body. At birth the head extends straight forward, as in most fish. It will assume its characteristic long-faced equine appearance later on, and the abdomen will curve in. The sea horse is propelled by a transparent dorsal fin; as it swims vertically, it beats its pectoral fins, which look like bristly ears, in unison. When tired, the sea horse wraps its prehensile tail around an aquatic plant and lets itself be rocked gently by the movement of the water; it is protected by a grillwork of cartilaginous armor topped by spines that run all the way down its back.

Sea horses are never more than six inches long; they feed on small animals and plankton. Despite their unusual appearance, they belong to the class of bony fishes. Their twenty species live in the coastal waters of western Europe, from the Mediterranean to the Black and North seas, with many additional species along the coasts of North America.

Despite the bizarre appearance of the sea horse, it ▲ is well camouflaged in its environment.

◀ Unique in the animal kingdom, the male sea horse becomes "pregnant": several females lay their eggs in his brooding pouch, which swells as the larvae develop. After a gestation period of six to eight weeks, the young are expelled into the water.

The Dolphin

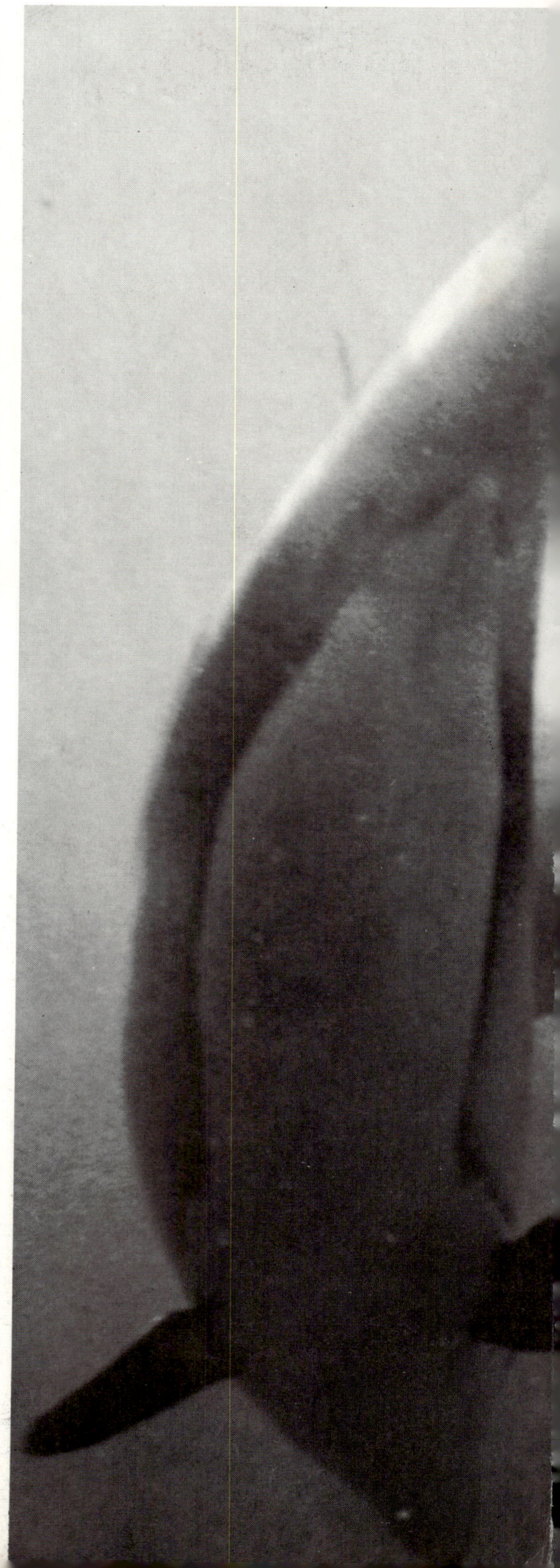

It was recently disclosed that the United States Navy was using dolphins in research. Previously regarded simply as a whale more amusing than most, dolphins were known best to ship passengers who delighted to watch them frolic in the vessel's wake. Our increasing knowledge shows that the dolphin may well prove to be a most valuable animal to man. Scientists have devoted major studies to it; Robert Merle's *Day of the Dolphin* is an informed—and disquieting—novel on the subject.

There are several species of dolphins, most of them ranging in all the world's seas. All belong to the toothed whales; they live in groups of 10 to 150, have two hundred tiny dagger-like teeth, swim rapidly, are extremely agile, and have a voracious appetite satisfied by underwater animals. One of the dolphins is the killer whale, which is eighteen to thirty feet long, and hunts seals and other dolphins. The most famous of the dolphins, the one involved in current research, is the common dolphin, which inhabits all the seas of the Northern Hemisphere except for arctic waters. It is seven to eight and a half feet long, weighs 330 pounds, and its head can be as much as a quarter of its total body length.

Sailors have for a long time told intriguing tales of mutual aid among dolphins, in which the animals' high intelligence always seems to be a factor. Specialists have gone much further: two American zoologists, T. G. Lang and H. A. P. Smith, were able to establish, with the use of tape recordings, that dolphins speak a common multiform language articulated in twitterings and shrill whistles. Like many of their colleagues, especially John C. Lilly, another American researcher, they have begun experiments that indicate the possibility of verbal communication between men and dolphins. Another scientist, H. Chapin, discovered that these curious animals, lacking a sense of smell and afflicted with very poor eyesight, have a radar system like that of bats. They emit sounds, which are bounced back by obstacles; the dolphin perceives the

The careful common dolphin female (*Delphinus delphis*) leaves the open sea before dropping her young, seeking the mouth of an inlet or bay, where she will be better sheltered. ▶

Opposite:
A usual mammalian birth, head first, would be fatal to the dolphin calf—it would drown. The tail appears first, and the animal backs into the world.

The females of the herd surround the pregnant dolphin, warding off all dangers while the mother swims slowly and gives birth to her calf. ▼

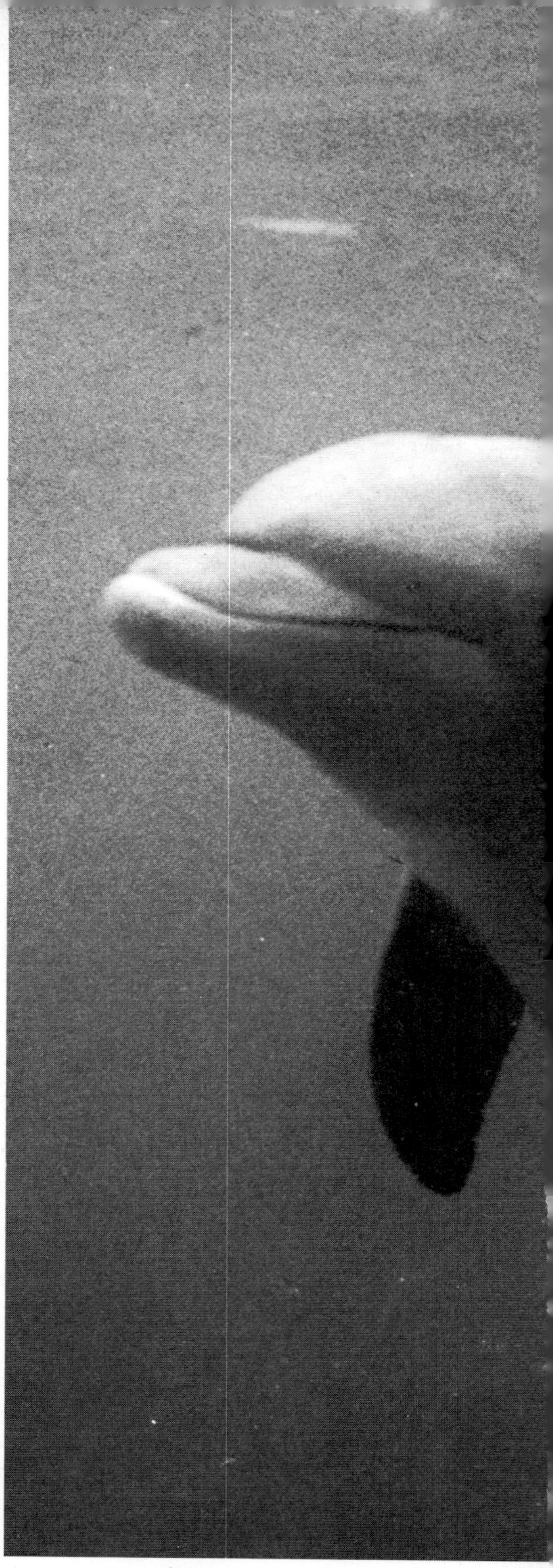

◀ The horizontal caudal fin does not interfere with the birth process since it is still soft. As soon as the young dolphin has entirely emerged from the mother's body, the fin stretches out and hardens and the calf is capable of swimming. Weighing twenty to thirty-five pounds at birth, the calf will double its weight in three months, thanks to its mother's milk, which is very rich in fats and proteins.

echo and heads straight for its source. Perhaps it is this characteristic that interests the American military.

It has been revealed by dissection that dolphins have two characteristics in common with bovines: a four-chambered stomach and blood of similar molecular composition.

In freedom, dolphins live in groups led by a bull surrounded by several females, to whom he pays attention only during mating. The calf, born after a gestation period of a year, monopolizes the mother's attention for the eighteen months it nurses. There is an interval of two or three years between births. The female raises her offspring, teaches it to hunt for food, and plays in its games. When it is weaned she leaves it on its own, to become integrated into the herd and travel with it.

Before they became the object of the attention of scientists and military personnel, dolphins in captivity were the joy of animal trainers. Dolphins obey all commands, juggle amazingly well, and seem to enjoy towing water-skiers. Their owners claim that they even invent their own tricks. Of all the animals, this intelligent cetacean may have the most surprises in store for us.

▲
Nursings alternate with breathing exercises. The calf slides its head under its mother's abdomen, grasps a teat between its tongue and lips, and sucks her milk. Nursing is repeated every ten, twenty, or thirty minutes; in between nursings, the calf surfaces for air.

Frigate Birds, Gannets, and Petrels

Above:
One of the petrels is the strictly monogamous fulmar (*Fulmarus glacialis*). It keeps the same mate for years. Male and female take turns brooding their single egg. The baby fulmar is born after forty to sixty days, and remains under the direct protection of its parents for only two weeks. It feeds from the nest for one to two months more, and is sexually mature only at the age of six.

Opposite, above:
Snow white with black wings, gannets (*Sula bassana*) are expert divers. They are found in the North Atlantic as far north as the Arctic Circle. They often spend the winter in Mediterranean Spain or North Africa. They line up their nests in close formation on jagged cliffs over the sea.

Opposite, below: The magnificent frigate bird (*Fregata magnificens*) nests from the Bahamas to the Cape Verde Islands. At mating time, the male puffs up his scarlet chest sac so that his beak rests on it. It is the male that builds the nest and broods the eggs.

No words can describe the majestic freedom of frigate birds, gannets, and petrels in the air. Their ease of movement in the skies is equaled only by their clumsiness on the ground, where they become tangled in their wings and are poorly supported by their frail legs.

The frigate bird, the open-sea equivalent of the eagle, inhabits tropical coastal areas. It is long (forty inches) and light (three and a half pounds), and has a huge wingspan of seven and a half feet. Almost incapable of taking flight from the surface of either ground or water, it perches on a tree branch or rock and takes off by diving into the air. It hunts for food only on the high seas, swooping down to catch fish, preferably flying fish; it strikes quickly, seizes the victim at wave level in its hooked beak, and immediately rises again vertically, swallowing its prey. Its fish diet is supplemented by sea snails, octopuses, and crabs, as well as young sea turtles, which it kills by the thousands.

Always remaining near the sea, frigate birds live and nest in colonies. The male builds the nest high in a tree and broods the single egg laid by the female. Mother and father both feed the baby, and both must protect it, as frigate birds are cannibalistic.

The gannet, less of a flier but as skilled at sea, is the size of a goose. It flies over tropical or temperate seas, going far out over the ocean. When it spots its prey it dives thirty or forty yards, wings folded back. To seize the victim, the gannet readily slips underwater in one flashing motion, its nasal passages sealed closed and protected against the shock of impact by air-filled cells under its skin, and surfaces triumphantly with the fish in its beak. Gannets live in giant communities, nesting in cliffs; they supply much of the guano for industry. The young must learn early to fend for themselves, as their parents do not train them. They are not sexually mature until the age of four.

Petrels, which look much like gulls, live in the endless expanses of ocean from the Arctic to the Antarctic. They fly tirelessly at fifty miles an hour, rarely moving their wings; they can cover tens of thousands of miles in several weeks. When released from an airplane in an experiment, 75 percent were observed to return without hesitation to their home port. How do they orient themselves? This mystery has not yet been solved.

Seals

Seals live in all the world's seas. They are ancient, highly evolved carnivores that seem to be related to the weasel family. They like to remain near land and, even on long excursions, prefer traveling along the coast rather than on the open sea.

Fast and skillful, they hurl their slender torpedo-like bodies through the water, agitating their limbs like fins. On land, however, they are heavy and clumsy. Seals are protected from the cold by a thick layer of fat; their fat also lessens the pain of the bites inflicted when the bull males are fighting over the females. The bull is generally surrounded by a veritable harem.

With auricular and nasal passages well closed, seals can remain underwater for some time—the record for Weddell's seal is thirty-one minutes at 1,950 feet. Usually, however, the animal paces itself, remaining submerged no more than ten minutes at a time. Seals are intelligent, have excellent hearing, and, depending on the species, barks like a dog or moos like a cow.

At the end of a ten-month gestation period, the female gives birth to one or two relatively large calves that are already protected by a furry coat. They are able to swim immediately after birth; they are rapidly weaned, and begin hunting for octopuses, shellfish, crabs, and other water animals. They fast during the mating season, when they congregate on islands.

Seals naturally live twenty-five to forty years, but the fur of the easily slaughtered

▲ The whiskered and waggish common seal of the North Hemisphere.

Northern fur seals on their breeding grounds. ▶

▲
The common seal (*Phoca vitulina*) sunbathes on ice floes, and prefers to remain in coastal areas.

Above left:
When there are two offspring, the common seal mother faces a terrible dilemma. Each calf swims away from its mother, although it cannot yet fend for itself. Whichever one is not followed by the mother perishes, pathetically calling out for her.

young brings high prices. Protective measures have been necessary to halt the massacre, especially in Canada.

The common seal lives in the Northern Hemisphere and can be found in temperate as well as polar latitudes. It is a true seal, as contrasted with the eared seals, such as sea lions. On land, the common seal has a difficult time moving its 165 pounds and six-and-a-half-foot length; but it is a marvelous swimmer. It hunts for food mostly at night; during the day it dozes in the sand or sunbathes on the ice.

The female common seal is faced with a strange dilemma. Her newborn swims away from her almost immediately after birth, although it cannot yet survive on its own. The mother must therefore follow her calf—but when there are two offspring, she can follow only one, and the other necessarily perishes.

Penguins

▲ Emperor penguins (*Aptenodytes forsteri*) have an exemplary social life. They never fight, and are characterized in all they do by great dignity.

Although they are birds, for perhaps a hundred million years penguins have been unable to fly, and have had to remain content with swimming and diving. A human-sized penguin fossil was discovered in Patagonia that dated back to the Tertiary Period, some thirty million years ago. Present-day penguins are usually far smaller. All penguin populations originated in the Antarctic, and today are scattered along the coasts of Australia, New Zealand, southern Africa, South America, and the Galapagos Islands.

These highly developed aquatic birds use their wing stumps as fins; their feathers have a scalelike texture. They propel themselves in the water by rowing vigorously, steering with their webbed paws. When swimming near the surface, only the head, neck, and narrow back remain above the water. At the rate of 120 and sometimes even 200 wing strokes a minute, they can attain a speed of twenty-four miles an hour; they dive at thirty feet per second.

Like dolphins, penguins must surface at regular intervals to breathe. They can leap forty to fifty inches in the air, clearing rocks this way.

Penguins feed solely on fish and sea animals; since they court, mate and brood on land, they must fast during this time. They crowd into their traditional brooding grounds by the tens of thousands.

Some penguins brood their eggs—one from each mating—in a pouch located at the base of the stomach. The female begins the brooding, then slides the half-incubated egg into the male's pouch and returns to the sea. Many of the brooding males, huddled together for sixty-three days, perish in ice storms. Soon after hatching, the females come back to their young. Now the exhausted males return to the sea, though many never reach it.

◀ It is believed that when the giant southern continent of Gondwanaland broke apart, penguins were carried along with the southern portion, now Antarctica. These survivors of an ancient stock are the most highly evolved of the aquatic birds.

King penguins (*Aptenodytes patagonicus*), which stand forty inches tall, gather in giant communities. Each couple's single egg is brooded during the night portion of the polar year. The outside temperature is then −30° F. to −50° F., but in the brooding pouch it is 104° F. When the young hatch, the adults build a wall to shelter ▼ them from ice storms.

The Polar Bear

The far north of the globe, the ice drifts of the North Pole, is the home of the polar bear. This animal shares with the Alaska brown bear the title of world's largest carnivore: it is nine and a half feet long and weighs up to 1,540 pounds. Always alone, it defies the ice storms of its snowy domain, since it is protected from the harshest temperatures by a thick layer of fat and a heavy coat of silvery-white fur that turns yellowish with age. In extreme weather it digs a hole in the snow for shelter.

The polar bear hunts mainly seals, approaching them from downwind. The bear silently swims up to its prey lying on the ice, dozing in the sun; it cuts off any possible retreat, or corners it underwater. Against young seals, which are clumsy on land, the polar bear is sure of a triumph without danger or glory. It also catches fish and sea birds, and its skill is so great that it rarely goes hungry. In autumn it goes on a plant diet, gathering berries, fruit, and grass on the tundra. The polar bear is as

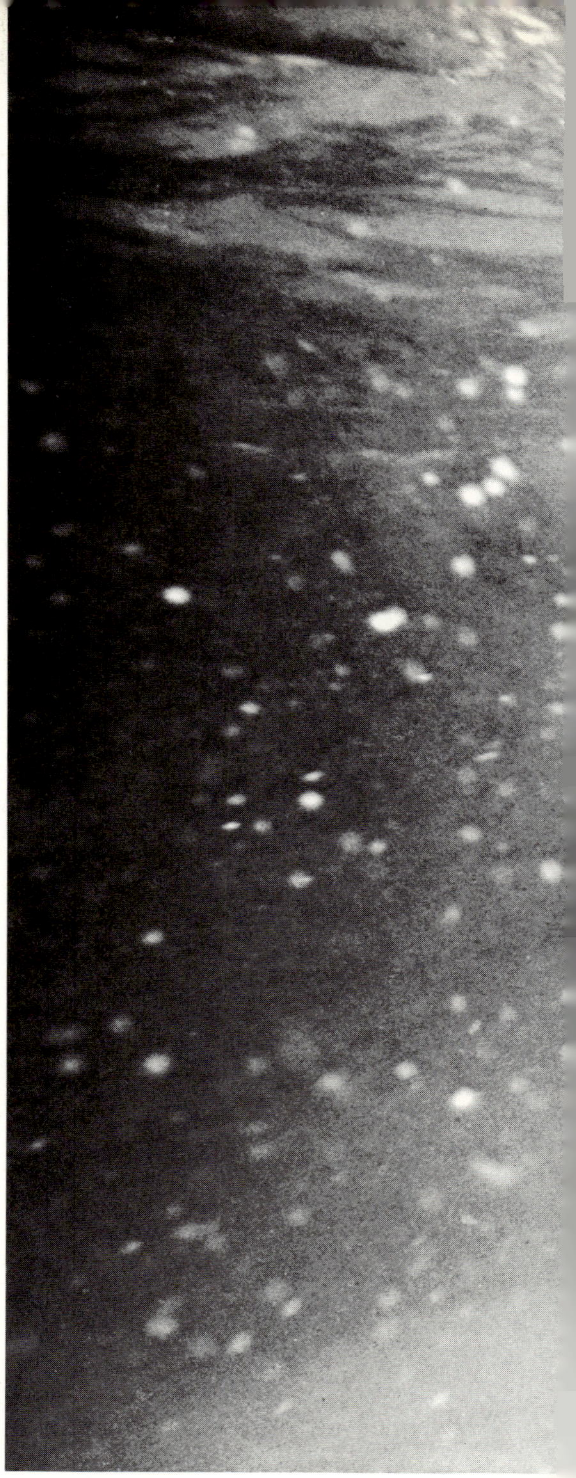

▲
The polar bear (*Thalassarctos maritimus*), which is protected against the bitter cold by its fat and its fur, swims rapidly and silently as it hunts seals and fish.

◀ Curious, droll, helpless, polar bear cubs live under their mother's care for a long period of time. She nurses them for several months, then trains them; they do not leave her for three years.

The cub lives its sweetest days curled up in its mother's fur. ▶

curious as a child, and on its long solitary meanderings it observes everything attentively, often sitting quietly on ice floes to examine its surroundings.

Such placidity in the polar bear is easily understood, for it is threatened by no animal—except man. The Eskimos and northern Asiatic peoples hunt the polar bear relentlessly, and its numbers are becoming fewer.

In December, just before giving birth, the female digs a den in the snow, igloo-fashion. She shelters her lair beneath a hugh pile of snow or builds an incline behind it as a windbreak. A ten-foot corridor leads to a chamber four and a half feet on a side and four feet high, sealed off by a block of ice or snow. The walls and floor melt from the heat of the bear's breath, but immediately freeze up again.

At the end of the eight-month gestation the mother gives birth to one to three cubs—generally two. The young are the size of rats, and are blind and weak. The mother immediately draws them to her breasts between her front legs. At four weeks, fuzzy fur grows in, and at six weeks the cubs leave their ice-home for the first time. When they are a year old they can fend for themselves, but the mother keeps them with her until they are three years old, apparently sending them alone into their icy world only when she is expecting another litter.

Sharks

During World War II, the United States Navy had to face the problem of sharks. Even if pilots shot down over water managed to survive the impact, sharks threatened them with mortal danger. Even chemical products could not repel them. Today the sporting industry has again taken up research: deepsea diving endangers thousands of people each year. The sponge and coral fishermen of tropical countries have had to face these dangerous flesh-eaters for hundreds of years. A solution is all the more difficult to find because the behavior of sharks follows no apparent pattern: why do the same species of shark attack man at one latitude and not at another? Is it only the individuals that have once tasted man's blood that become man-eaters? These questions are still unanswered. Electronic systems of defense were being tested when a German scientist, Hans Hass, suggested an empirical defense: high-pitched screams.

One fact alone is definite, and that is the instinctive taste for murder of most of the 250 species scattered throughout the oceans. Stewart Springer, an American marine biologist, observed while studying a pregnant tiger shark that the fetuses were trying to devour each other.

The appearance and behavior of the many shark species vary greatly, except for three characteristics that all share: jaws lined with several rows of teeth always growing backward; respiration by means of open gill slits; and a torpedo-shaped

body, propelled by a powerful caudal fin, that allows them to cleave the waters. It is hard to remember that the small dogfish of the North Sea, only two feet long, is related to the terrible sixteen-to-thirty-foot blue shark of the tropical and subtropical seas—and to the dread hammerhead shark, and to two giants: the whale shark, which can be sixty feet long, and the basking shark, which is forty-five feet long and weighs four and a half tons. Paradoxically, these last two are harmless; they feed on plankton, and cause damage only if they accidentally ram a boat.

◀ Like murderous torpedos, sharks (Selachii) dive to surprise and attack their victims. They can smell the blood of wounded animals at great distances, and can recognize the behavior of a sick fish. Their attacks on man follow no apparent pattern.

Its enormous head, with an eye and nostril at each side, gives the dread hammerhead shark (*Sphyrna zygaena*) a bizarre appearance. The larger its size, the more aggressive it is. ▶

▲ It is too late to flee: this diver faces a shark swimming straight for him.

The greatest depths are the hunting grounds for several shark species; others swim near the surface. ▶

Biomes and Habitats

1. Plains, Savannahs, Prairies

As one moves away from the equator, rain forest gives way to grassland. Initially, trees dot the open spaces, following water courses. Then the trees become progressively sparser, giving way to bushes no taller than the surrounding sea of grass.

2. Deserts and Semi-Arid Lands

Where there is drought for ten months of the year, the grassland thins out and gives way to semi-arid and desert land. Little or no vegetation grows, and roots no longer bind the soil together. The wind creates sand dunes, which can be as high as nine hundred feet, and totally sandy wastes cover vast areas, such as the Great Western Erg in the Sahara, which is more than thirty thousand miles square.

3. Tropical and Subtropical Forests

Dense, interwoven vegetation flourishes year-round in the forests of tropical and subtropical regions. The air is heavy with moisture, and little light penetrates the labyrinth of greenery.

4. Evergreen Forests

In northern regions, trees with cones and needle-like leaves make up dense evergreen forests. Firs, pines, and spruces are the most common conifers. These trees can grow as tall as 150 to 180 feet. A dense carpet of needles form a floor that does not easily decay and prevents the growth of underbrush.

5. High Mountains

These are the snow-covered peaks that rise over 8,000 feet. Only mosses and lichen grow above the timberline, which is usually at 5,550 feet.

6. Oak and Beech Forests

In addition to oaks and beeches, the dominant species, maples, elms, birches, sycamores, hickories, and alders are found in deciduous forests.

7. Open Woods and Scrubland

The vegetation is open, with small stands of trees interrupted by sun-filled glades. A great variety of animals inhabit these partially cleared lands, seeking shelter and hibernating in the deeper parts of the woods.

8. Fields and Meadows

This is the home of many species of insects and rodents. The vegetation is grassy, interspersed with stunted trees, briars, bushes, and many wild flowers.

9. Rocks and Caves

Natural rocky areas and caves as well as man-made ruins are the habitat of rock-nesting birds and bats.

10. Marshes and Tundra

Marshes are poorly drained lowlands overrun by shallow stagnant waters or gradually rising alluvial deposits. Tundras, which are often swampy in summer, are the treeless regions of the far north above the coniferous forests.

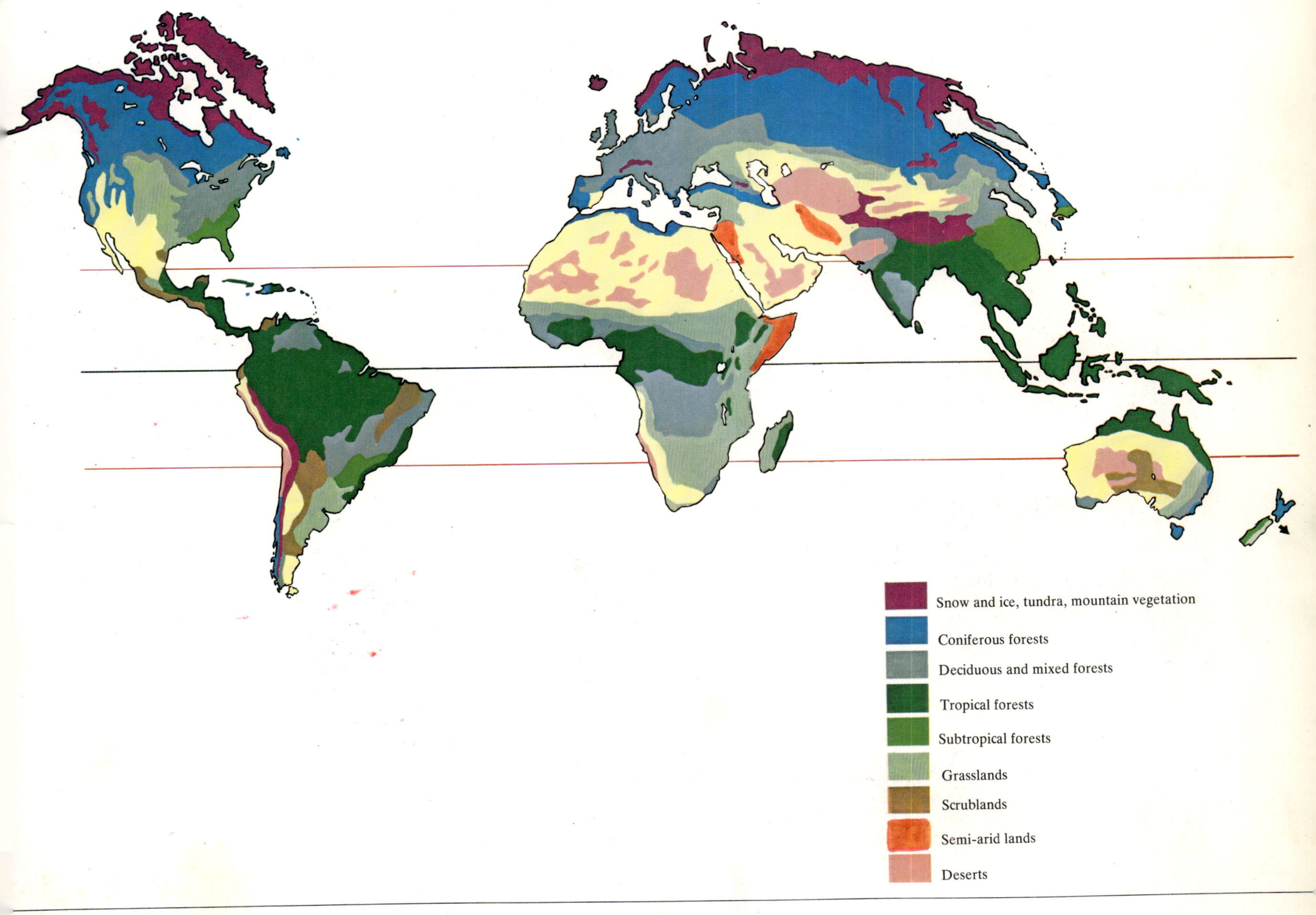

11. Rivers and Oceans

Many animals are adapted to live in the water, both fresh and salt. Penetrated by the bright rays of the sun, coastal waters are a field of active photosynthesis that permits a rich explosion of plant und animal life. Continental waters are also a rich environment, and provide a stopping place for migrating birds.

Index

Photograph Credits

Afrique Photo, Paris: 23 (above), 23 (below), 29 (above), 30 (above), 42–43, 46 (above), 97 (below), 98 (above), 237 (below), 290–291, 290 (below). *Bavaria-Verlag, Gauting:* 41 (below), 53 (right), 64 (above), 80, 84 (left), 92 (above), 94 (left), 114 (above), 114 (below), 121 (below, right), 125 (left), 125 (right), 132 (left), 132–133, 133 (below), 134 (left), 140 (left), 141 (below), 148 (left), 148 (right), 148–149, 153 (below), 160, 161 (above, left), 164 (below), 167 (above), 171 (above), 174 (above), 175, 198 (above), 201 (above), 201 (below), 203 (below), 204–205, 205 (right), 208 (below), 212, 213 (left), 213 (right), 221 (above, right), 224 (right), 225 (left), 225 (above, right), 229 (above), 229 (below), 231 (below), 232 (above), 233 (left), 236–237, 240 (above), 240 (below), 244 (left), 248–249, 249 (above, left), 249 (right), 249 (below), 252 (left), 252–253, 262 (left), 264 (above), 268, 269 (above), 276 (left), 280, 281 (left), 285 (below), 305 (right). *Black Star, New York/Photos Flip Schulke:* 310 (above), 310 (below), 311, 312 (left), 312–313. *Arthur Christiansen, Copenhagen:* 144, 145. *Comet-Photo AG, Zürich:* 78 (left). *Dpa, Frankfurt:* 21, 39 (below, left), 48, 53 (left), 76 (below), 81, 84–85, 94–95, 109 (above), 110, 113, 116–117, 118 (left), 118–119, 119 (above), 121 (below, left), 181, 263 (below), 285 (above, right), 297 (right). *Expéditions polaires françaises, Paris/Photos Kirchner:* 306–307. *Foto-Garbrecht, Hamburg:* 55 (above), 87 (above, right). *Sven Gillsäter, Tiofoto, Stockholm:* 305 (above, left), 305 (below). *Globe Photos, Inc., New York:* 298, 298–299, 299 (above), 300 (above), 300 (below), 300–301. *Gerhard Gronefeld, Munich:* 8–9, 37. *Martin Grossmann, Berlin:* 266–267, 270. *Jürgen Klages, Zurich:* 49, 308. *Aldo Margiocco Campomorone:* 10 (below), 20 (below, right), 54, 55 (center), 55 (below), 56, 58 (above), 59, 82, 96 (above), 97 (above), 108 (below), 122 (below), 134 (right), 136 (below), 137, 139 (below, left), 150 (below), 170, 172 (above, left), 172 (center), 172–173, 173 (below), 187 (below), 189 (center, right), 194 (below), 196, 197, 200 (below), 206, 207, 214 (center), 214 (below), 219 (above, right), 219 (below), 226 (above, left), 226 (above, right), 226 (below, right), 235 (above), 240 (center), 246 (below), 265, 275, 282. *Dr. H.-J. Matthäi, Lenggries:* 128, 130 (above, left), 130 (below), 131 (above), 131 (below). *Norwegian Tourist Office/Photo P. A. Röstad, Oslo:* 216 (right). *Tierbilder Okapia, Frankfurt:* 19 (above), 30 (below, right), 30 (below, left), 34 (center), 34 (below), 64 (below), 65, 68, 72 (below), 76 (below), 83 (below), 86 (center), 88, 89 (above), 89 (below), 92 (below), 98 (below), 101, 106 (below, right), 129 (above, right), 129 (below, right), 139 (below, right), 140, 151, 175, 176 (left), 184 (center), 208 (above), 209 (above), 209 (below), 232 (below), 236 (below), 248 (above), 263 (above), 274 (above), 283 (below), 284, 285 (center, left). *Klaus Paysan, Stuttgart:* 11 (above), 11 (below), 12 (below), 14 (below), 15, 16, 17, 18 (above, left), 18 (above, right), 18 (center), 18 (below), 19 (below), 20 (below, left), 24 (below, left), 24 (below, right), 25, 26 (above, left), 26 (below), 28–29, 31 (below, left), 31 (right), 32 (below), 33 (above), 33 (below), 34 (above), 35 (above), 35 (below), 36, 38 (below, left), 38–39, 39 (below, right), 40, 43 (above), 44, 45 (left), 47 (above), 47 (below, left), 47 (below, right), 52, 57 (above), 57 (below), 58 (center), 58 (below), 62 (left), 63 (above), 63 (below), 66–67, 85 (below), 86 (below, right), 90 (left), 90 (right), 91 (above), 91 (below, left), 91 (below, right), 96 (below), 102–103, 111 (below, right), 115 (right), 120 (left), 124 (left), 124 (right), 126–127, 146–147, 150 (left), 154 (above), 156 (above), 156 (below), 159 (below, right), 161 (below), 166 (above), 167 (below), 189 (below, right), 199 (above), 199 (below), 202 (below), 215 (above), 220 (below, center), 220 (below, right), 224 (left), 225 (right), 236 (above), 238–239, 243, 250 (below), 257 (center, left), 288, 289 (below), 286 (above), 286 (below), 287 (above), 295 (above, right), 304 (below). *Hans Pfletschinger, Reichenbach/Fils:* 116 (below, left), 121 (above), 155, 157, 159 (above), 165, 173 (above), 184 (above), 186 (above, left), 186 (above, right), 187 (above), 188 (above, left), 188 (below), 189 (above, left), 189 (above, right), 189 (below, left), 192 (above), 192 (below), 193 (above), 193 (below), 194 (above), 194 (center), 195 (above), 195 (below), 214 (above), 215 (below), 218–219, 220 (below, left), 221 (below, right), 222 (above), 222 (below), 223 (above), 226 (below, left), 227, 230, 234, 235 (center), 235 (below), 241 (above), 251 (above), 251 (below), 256, 259 (below), 289 (above). *Fritz Pölking, Greven:* 153 (above), 166 (below), 168–169, 168 (below), 169 (below), 180 (above), 191 (above), 191 (below), 202 (center), 228 (below), 257 (above), 264 (below), 277 (above), 277 (below). *Georg Quedens, Norddorf:* 104–105, 152 (below), 162 (above), 174 (below, right), 180 (below, left), 200 (above), 203 (below), 271 (above), 271 (below). *Rapho, Paris:* 13, 27, 38 (below, right), 45 (right), 111 (above), 142 (left), 143, 148 (below), 182–183, 184 (below), 223 (below), 287 (below), 292 (below), 294, 295 (above, left), 295 (above, right), 297 (left), 309 (below). *Réalités, Paris:* 24 (above), 50–51, 62–63. *Günter Reitz, Hannover:* 28 (below), 32 (above), 46 (below). *Roebild, Frankfurt:* 35 (below), 48, 68–69, 93 (below), 107 (below), 217, 231 (above, right), 244–245, 286 (above). *Eugen Schuhmacher, Grünwald, Munich:* 303 (below). *Emil Schultess, Forch/Zurich:* 283 (above), 292–293, 293 (below). *Stern Hamburg, Photos Scheler:* 60. *Betzler:* 61. *Archiv:* 14 (above), 74, 272, 273 (below). *Tass, Moscow:* 40 (above). *Time-Life Picture Agency, New York:* 70–71, 72–73, 75, 78–79, 308–309. *UPI, Frankfurt:* 306 (above). *Widerøe's Flyveselskap, Oslo:* 260–261. *ZFA, Düsseldorf:* 2–3, 12 (above), 20 (above), 22, 60 (above), 83 (above), 86 (above, right), 86 (below, left), 87 (below), 93 (above), 99, 100 (above), 100 (below), 106 (above), 106 (below, left), 107 (above), 108 (above), 111 (below, left), 112, 114–115, 115 (below), 119 (below), 121 (above), 122 (below, left), 123, 129 (left), 135 (above), 135 (below), 136 (above), 138–139, 142 (right), 152 (above), 154 (below), 158 (above), 159 (below, left), 161 (above, right), 162 (below), 163, 164 (above), 171 (below), 174 (below, left), 176 (right), 177, 185, 190, 198 (below), 210–211, 216 (above), 216 (below), 220 (above), 221 (left), 228 (above), 231 (above, left), 233 (right), 241 (below), 242, 247, 254–255, 258 (left), 258 (right), 259 (above), 273 (above), 274 (below), 278–279, 281 (right), 291 (below), 295 (center), 296, 306 (below). *Zoölogisch Laboratorium Universiteit van Amsterdam/Photos Dr. A. Kortlandt:* 77 (above), 77 (below).